An Introduction to Liquid Helium

AN INTRODUCTION TO LIQUID HELIUM

Second Edition

J. WILKS and D. S. BETTS

Fellow of Pembroke College
University of Oxford

Reader in Experimental Physics
University of Sussex

CLARENDON PRESS · OXFORD
1987

Oxford University Press, Walton Street, Oxford OX2 6DP
Oxford New York Toronto
Delhi Bombay Calcutta Madras Karachi
Petaling Jaya Singapore Hong Kong Tokyo
Nairobi Dar es Salaam Cape Town
Melbourne Auckland
and associated companies in
Beirut Berlin Ibadan Nicosia

Oxford is a trade mark of Oxford University Press

Published in the United States
by Oxford University Press, New York

First edition 1970
Second edition 1987

British Library Cataloguing in Publication Data
Wilks, J. (John)
An introduction to liquid helium.—2nd ed.
1. Superfluidity 2. Liquid helium
I. Title II. Betts, D. S.
546′.751 QC175.4
ISBN 0-19-851471-9

Library of Congress Cataloging in Publication Data
Wilks, J. (John), 1922–
An introduction to liquid helium.
Bibliography: p.
Includes index.
1. Liquid helium. I. Betts, D. S. (David Sheridan), 1936– . II. Title.
QC145.45.H4W55 1986 536′.56 86-16422
ISBN 0-19-851471-9

Typeset and printed in Northern Ireland by
The Universities Press (Belfast) Ltd. Belfast.

PREFACE TO THE SECOND EDITION

The first edition of this book by one of us was published in 1970 and designed as an abridged form of the larger book *The properties of liquid and solid helium* published in 1967. The subject has progressed considerably since the sixties, and the time has come to present a new edition. The original arrangement of chapters has been largely retained but for the removal or redistribution of the topics covered in the old Chapter IV, the renumbering of the old Chapters V to IX as new Chapters 4 to 8, and the insertion of a substantial new Chapter 9 on the superfluid phases of liquid ^{3}He. The discovery in 1972 of superfluidity in ^{3}He, and the subsequent development of our understanding of it, provides the largest continuous body of new material in this new edition. But investigations of many other aspects of the behaviour of liquid helium have progressed considerably in the last twenty years, and the whole of the first edition has had to be revised and updated. This will be most noticeable in sections on, for example, critical velocities, mixtures, and properties near the critical points and phase transition surfaces. But most sections have been brought up to date to a greater or lesser extent, and the reference list has been doubled in length, with the result that there has been a general expansion. Also the whole book has been reset, with 47 new figures out of a total of 101, so that the appearance and feel of the book are changed and, we hope, improved.

Oxford J. W.
Brighton D. S. B.
August 1986

CONTENTS

1
INTRODUCTION

Liquid helium exists in two isotopic forms, liquid ^{3}He and liquid ^{4}He. The most important characteristic of both liquids is that under their saturated vapour pressures they remain liquid down to the very lowest temperatures. We begin by indicating how this comes about.

1.1 The permanent liquids

The helium atom is spherically symmetrical, and smaller than that of any other element. The only binding forces in the liquid are van der Waals forces, which arise from the fluctuating polarization charges induced in the electron shells of adjacent atoms. These forces are weaker than in all other substances, so the critical and boiling points of helium (Table 1.1) are the lowest of all.

In most substances, the freezing point would be determined by a balance between the Van der Waals forces and the thermal energy. The attractive forces tend to pack the molecules into a regular crystalline lattice, while the thermal energy tends to disrupt an ordered arrangement. However, in liquid helium both the attractive forces and the thermal energy are extremely small. It thus comes about that another contribution to the balance of forces, usually negligible, must be taken into account. This is the effect of the so-called zero-point energy.

To a first approximation, we may regard an atom in the solid as a harmonic oscillator, and its ground-state energy is given by the Schrödinger equation as $\frac{1}{2}h\nu_0$, where ν_0 is the characteristic frequency and h is Planck's constant. Alternatively, we may picture an atom in the liquid as a free particle located in a small cage formed by the adjacent atoms; the energy of the lowest state is then $3h^2/8mV^{\frac{2}{3}}$, where m is the mass of the atom, and V the volume of the cage. In both cases the energy is lowered if the molar volume increases, hence the effect of this ground-state, or zero-point, energy is to increase the molar volume.

Figure 1.1 shows the potential energy of both liquid and solid ^{4}He as a function of molar volume, estimated from the known force fields of the helium atoms and their spatial arrangement in the liquid and solid, as shown by X-ray scattering measurements. In a normal liquid the stable

Table 1.1. The critical temperature T_c, and normal boiling point T_B, of liquid ^{3}He and ^{4}He

	^{3}He	^{4}He
T_c(K)	3.32	5.20
T_B(K)	3.19	4.21_5

state at absolute zero would be that with the lowest potential energy. In this case we would expect the solid to be stable with a molar volume of about 10 cm^3 (see Fig. 1.1), but when considering condensed helium we must include the effect of zero-point energy. Figure 1.1 shows an estimate given by F. London for zero-point energy, which he assumed to be approximately the same in both the solid and liquid phases. At absolute zero the total energy of either the liquid or solid phase is the sum of the zero-point energy and the respective potential energy. The total energy of each phase has a minimum value at a molar volume of about 30 cm^3, and at this volume the total energy of the liquid is actually less than that

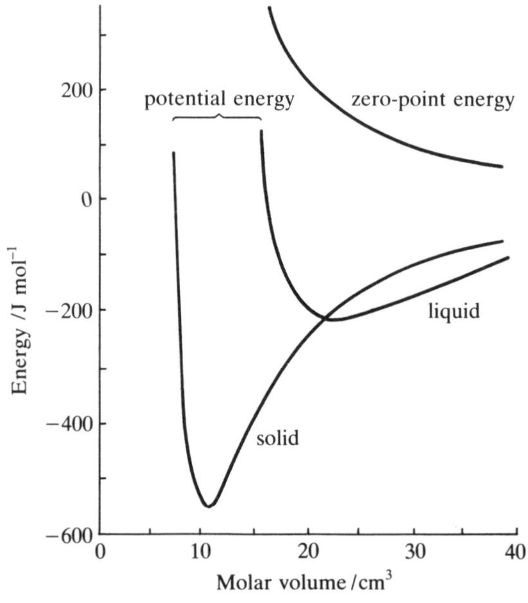

Fig. 1.1. Estimated values of the potential and zero-point energy of liquid and solid ^{4}He at absolute zero, as functions of the molar volume (London [1]).

of the solid. Hence the liquid remains the stable phase down to absolute zero, as is observed, with a molar volume of $28\,\mathrm{cm}^3$ in good agreement with the estimate. As the zero-point energy depends inversely on the mass of the atom, its effect is relatively greater in liquid ^{3}He, which has a molar volume of about $40\,\mathrm{cm}^3$ even though the atoms are the same size as those of ^{4}He. Although an exact calculation of the zero-point energy in the liquid and solid phases presents considerable difficulty, there is no doubt that helium remains liquid because the zero-point energy leads to a large molar volume.

Finally we note that both liquid ^{3}He and ^{4}He are colourless, with densities of about 0.07 and $0.14\,\mathrm{g\,cm^{-3}}$ respectively; and liquid ^{3}He has a much higher vapour pressure as the atoms are less tightly bound.

1.2 The melting curve

Liquid helium may be solidified by the application of pressure. The stable state of a system is that with the lowest free energy, defined in terms of the usual thermodynamic variables as

$$G = U - TS + pV.$$

At absolute zero, under the saturated vapour pressure, this condition reduces to the internal energy U being a minimum. However, if we apply pressure to the liquid and solid, the pV term contributing to the free energy is different for the two phases. The solid then becomes the stable phase at higher pressures.

The pressure required to solidify liquid ^{4}He is shown in Fig. 1.2. Below 1 K, this melting pressure becomes virtually independent of temperature, and the gradient of the melting curve decreases approximately as T^8. (We ignore for the moment an extremely small variation in the melting pressure which we mention again in Section 2.5.) According to the Clausius–Clapeyron equation

$$\left(\frac{\mathrm{d}p}{\mathrm{d}T}\right)_{\mathrm{m}} = \frac{\Delta S_{\mathrm{m}}}{\Delta V_{\mathrm{m}}},$$

where ΔS_{m} and ΔV_{m} are the entropy and volume changes on melting. Measurements of the volume change show that ΔV_{m} takes up a finite and constant value at low temperatures. Hence the entropy difference ΔS_{m} tends to zero, a result which is in accord with the Third Law of Thermodynamics. The entropies of both the solid and liquid vary approximately as T^3, and the entropy differences ΔS_{m} varying as T^8 falls off much more rapidly. Hence over an appreciable range of temperature, there is almost no heat of melting, and liquid and solid along the melting curve have virtually the same entropy. As we normally think of a liquid

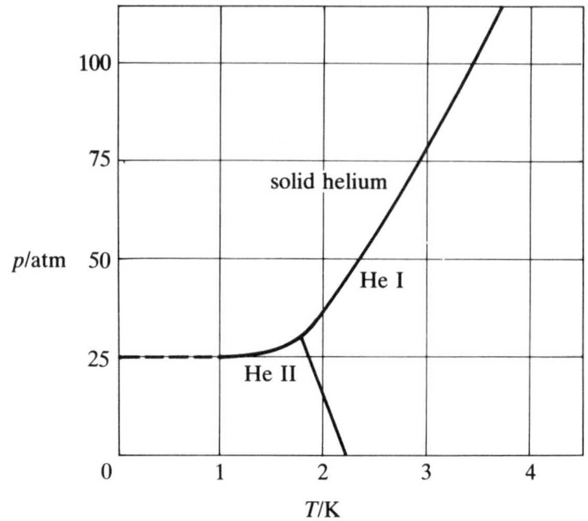

Fig. 1.2. The melting pressure and lambda line of liquid ^{4}He (Swenson [2]).

as a more disordered phase than a solid, it is not surprising that the liquid exhibits unusual properties.

1.3 Liquid ^{3}He and liquid ^{4}He

At first sight, liquid ^{3}He and liquid ^{4}He are not very different from the other inert gas liquids, except for their large molar volume. In fact, using the simple kinetic theory of gases we can account, with moderate success, for the thermal and kinetic properties of the liquids near to their boiling points. However, at lower temperatures, both liquids show very marked deviations from this quasi-classical behaviour.

In liquid ^{4}He the change in behaviour is marked by a very characteristic anomaly in the specific heat, the so-called lambda transition (Fig. 1.3(a)). The specific heat rises to a very high value at the lambda temperature of 2.17 K, and then falls off rapidly. Immediately below this transition temperature the liquid becomes *superfluid,* that is, it can flow through very fine channels with apparently no viscosity at all! In addition it exhibits a whole range of other unique properties.

The properties of liquid ^{3}He above 1 K are quite similar to those of liquid ^{4}He near the boiling point, as we might expect since the two atoms have electron shells which are almost identical. Yet in the temperature range all the way down to 3 mK liquid ^{3}He behaves quite differently from liquid ^{4}He and also from the behaviour expected for classical liquids. Finally, at about 2.6 mK, liquid ^{3}He undergoes a transition to a superfluid

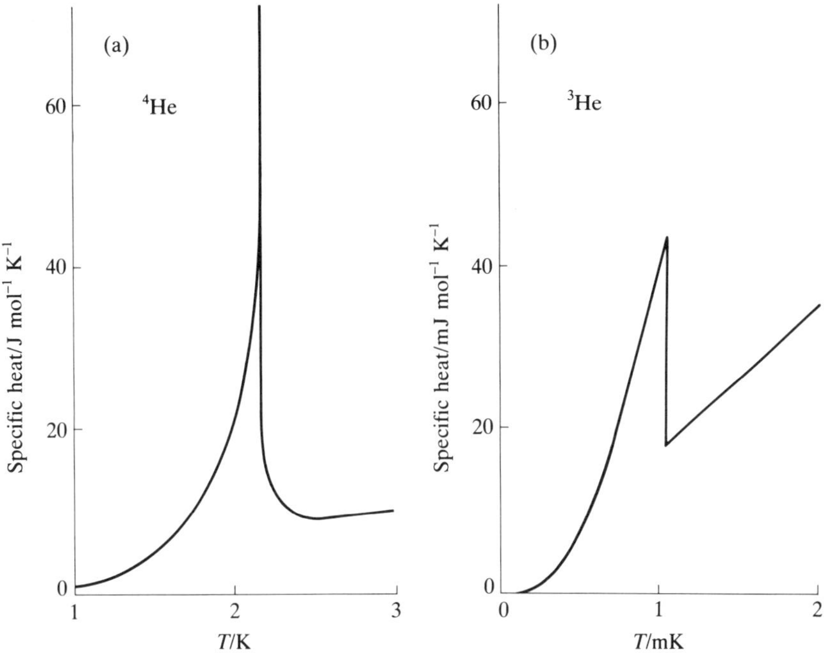

Fig. 1.3. The specific heats of (a) ^{4}He and (b) ^{3}He showing the anomalies associated with their respective phase transitions. Note that the scales in (a) and (b) are quite different, being a factor of 10^3 lower for ^{3}He. ^{4}He data is from Atkins [3]; ^{3}He data is from Alvesalo *et al.* [4].

phase, the transition being reflected in an abrupt increase in the specific heat (Fig. 1.3(b)). We now go on to describe 'normal' ^{3}He down to 2.6 mK, superfluid ^{4}He below 2.17 K and superfluid ^{3}He below 2.6 mK. We shall see that normal ^{3}He is governed by Fermi statistics and ^{4}He by Bose statistics. The superfluid transition in ^{3}He is associated with the formation of ^{3}He pairs which act as bosons.

2
NORMAL LIQUID ^{3}He

Above 1 K, liquid ^{3}He behaves somewhat like a dense classical gas, but at lower temperatures there is a transition to a quite different regime. We now outline some of the main features of this region.

2.1 The specific heat and susceptibility

Although we do not intend to give detailed descriptions of experimental arrangements, two features of work with liquid ^{3}He are particularly noteworthy. The transition to the low temperature region only becomes complete at temperatures below 0.05 K, and measurements down to at least a few millikelvins are necessary for a full study of the normal liquid. These are much lower temperatures than are found necessary in work on ^{4}He, and call for specialized techniques, both to attain the temperatures and to achieve thermal equilibrium within the experiment (see for example reference [5]). In addition the supply of ^{3}He was initially very restricted, and even today experiments are usually designed to use no more than a few cm^3 of liquid.

The specific heat has been measured by a number of workers, including Greywall [6], some of whose data are shown in Figs. 2.1 and 2.2. Below 100 mK the specific heat appears to be varying almost, but not quite linearly with temperature. The measurements taken below 0.1 K may be fitted to a form of curve derived from the theory discussed in Section 3.6

$$C_V/R = \gamma T + \Gamma T^3 \ln(T/\theta_c) \tag{2.1}$$

where γ, Γ and θ_c are volume-dependent positive constants. For a volume of 36.74 cm^3 mol^{-1}, $\gamma = 2.78$ K^{-1}, $\Gamma = 35.4$ K^{-3}, and $\theta_c = 0.458$ K and as T tends to zero C_V tends to the value 2.78 RT. Note that the second term on the right hand side of eqn (2.1) amounts to nearly 20 per cent of the main term at 0.1 K, and its exact form is important because of the effect on deduced values of the effective mass m^* of the ^{3}He atoms (see Chapter 3). There has consequently been debate about this matter which the precision of Greywall's work has done much to settle.

Figure 2.3 shows the magnetic susceptibility χ arising from the nuclear spin of $\frac{1}{2}$ associated with each ^{3}He atom, plotted in the form of χT vs T.

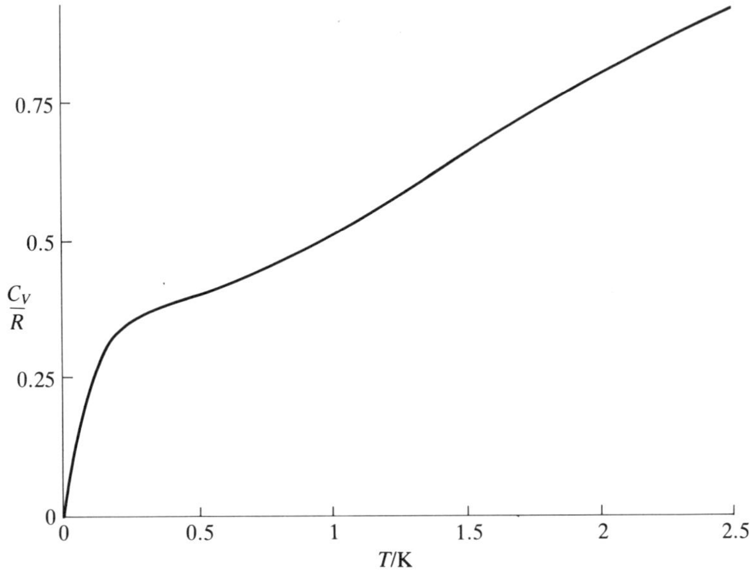

Fig. 2.1. The molar specific heat C_V (divided by R) of liquid ^{3}He at fixed volume $36.82\ \mathrm{cm^3\ mol^{-1}}$ corresponding to zero pressure at 0.1 K (after Greywall [6]).

At high temperatures the susceptibility varies as $1/T$ (i.e. χT is constant), but as the temperature is reduced the susceptibility falls below this value, and finally becomes independent of temperature (i.e. χT is proportional to T).

2.2 The transport coefficients

Above 1 K the coefficient of viscosity under the saturated vapour pressure is about 25 micropoise and approximately independent of temperature. At lower temperatures, however, it increases steeply as shown in Fig. 2.4, the results below 125 mK being best described (Bertinat *et al.* [8]) by the relation $1/\eta T^2 = 0.5450 - 1.594T$ or almost equivalently,

$$\eta T^2 = 1.83(1 - 2.9\, T)^{-1} \text{ micropoise K}^2.$$

Below 10 mK the second term in the bracket is scarcely perceptible compared to unity, and Parpia *et al.* [11] have observed a $1/T^2$ dependence down to the temperature of the superfluid transition, the viscosity at zero pressure rising to a value as high as 2 poise. Values at a pressure of 15.93 b are given in Fig. 2.5, which shows that the viscosity varies as $1/T^2$ and then falls sharply below the transition temperature T_c.

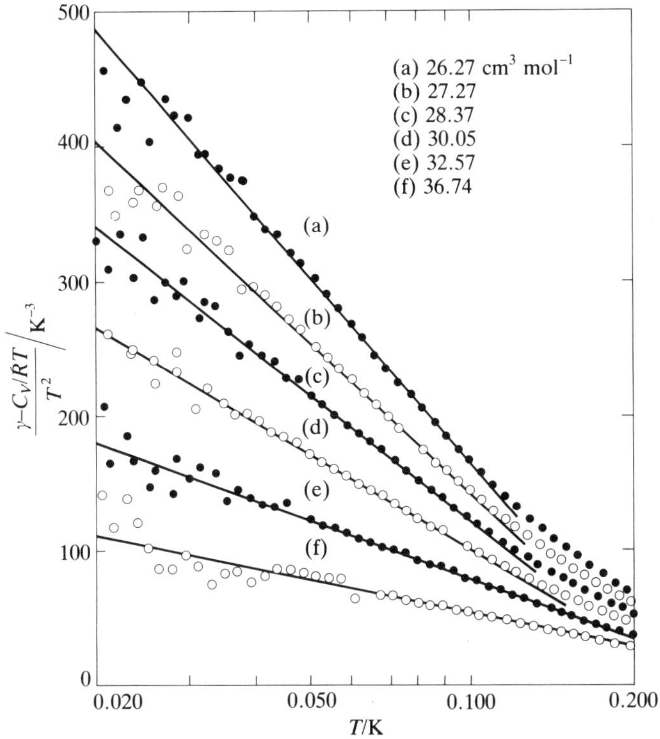

Fig. 2.2. The molar specific heat C_V of liquid ^{3}He at low temperatures and various molar volumes. The ordinates show the values of the quantity $(\gamma - C_V/RT)/T^2$ of eqn (2.1) in units of deg^{-3}. The horizontal temperature axis is logarithmic so the straight lines confirm the presence of the term $T^3 \ln(T/\theta_c)$ in eqn (2.1) (after Greywall [6]).

The behaviour of the thermal conductivity is shown in Fig. 2.6. It exhibits a minimum in the region of 0.2 K, and below 30 mK takes the limiting form

$$KT = 34.8(1 - 6.3\,T)^{-1}\,\text{erg s}^{-1}\,\text{cm}^{-1}$$

(Wheatley [14]). More recent data have been given by Greywall [15]. The coefficient of spin diffusion has also been measured [14] and at the lowest temperatures is given by

$$DT^2 = 1.44 \times 10^{-6}(1 - 2.9\,T)^{-1}\,\text{cm}^2\,\text{s}^{-1}\,\text{K}^2.$$

2.3 The perfect Fermi gas

It is now instructive to compare the properties of liquid ^{3}He in the low temperature region with those of a perfect Fermi gas. We therefore

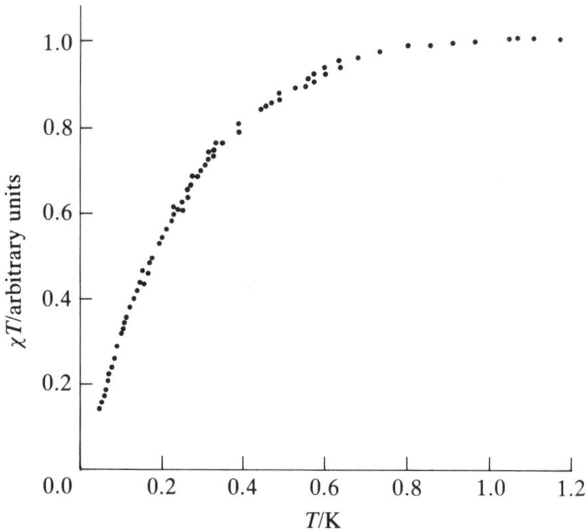

Fig. 2.3. The magnetic susceptibility χ of liquid ^{3}He under a pressure of 0.5 atm, as a function of the temperature. The product χT is plotted against T in arbitrary units (after Thomson *et al.* [7]).

gather together some expressions for the properties of a perfect gas of particles with spin $\frac{1}{2}$ obeying Fermi statistics in the degenerate low temperature region. We consider an assembly of N atoms of mass m in a volume V.

The distribution function giving the number of atoms with energy ϵ is given by the usual methods of statistical mechanics for an assembly of

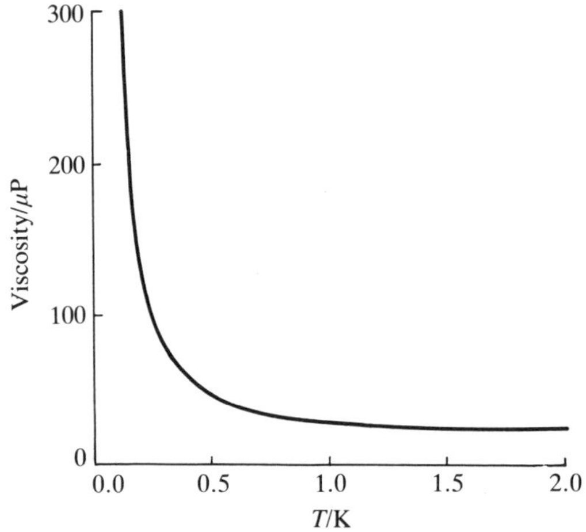

Fig. 2.4. The viscosity of liquid ^{3}He (after Betts *et al.* [9], [10]).

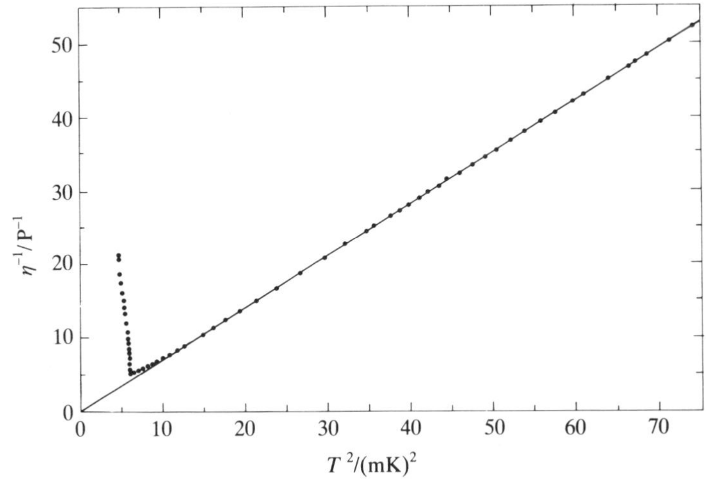

Fig. 2.5. The inverse of the viscosity of liquid ^{3}He at a pressure of 15.93 b as a function of the square of the temperature (after Parpia *et al.* [11]).

identical Fermi particles as

$$n(\epsilon) = \{e^{(\epsilon-\mu)/k_B T} + 1\}^{-1},$$

where k_B is Boltzmann's constant, and μ the chemical potential specified by the condition

$$\sum_{\epsilon} n(\epsilon) = N.$$

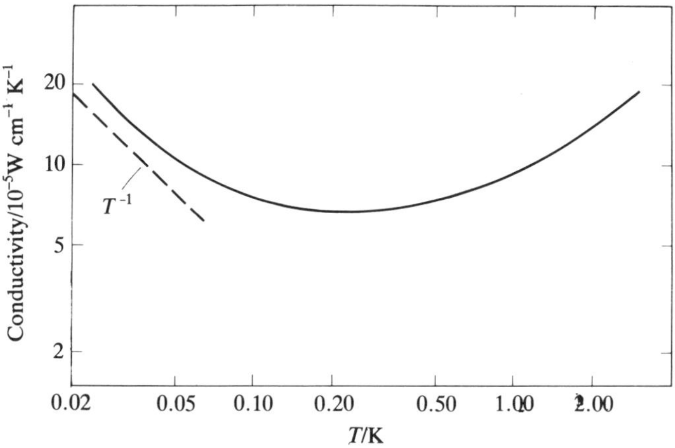

Fig. 2.6. The thermal conductivity of liquid ^{3}He (after Lee and Fairbank [12], and Anderson *et al.* [13]).

Note that μ is in general a function of temperature, but that at sufficiently low temperatures it approaches a constant value denoted by μ_0.

The Fermi surface. At temperature zero, all the energy levels are filled up to the level $\epsilon = \mu_0$. This energy is the *Fermi energy,* which has the value

$$\mu_0 = \frac{\pi^2\hbar^2}{2m}\left(\frac{3N}{\pi V}\right)^{\frac{2}{3}}.$$

The momentum associated with this energy level has the value

$$p_0 = \pi\hbar\left(\frac{3N}{\pi V}\right)^{\frac{1}{3}}. \tag{2.2}$$

The boundary of the filled levels (at temperature zero) in momentum space is termed the *Fermi surface,* and for an isotropic system takes the form of the surface of a sphere of radius p_0. In a *degenerate* Fermi gas (i.e. $k_B T \ll \mu_0$) the distribution function $n(\epsilon)$ does not differ greatly from its value at absolute zero. In fact $n(\epsilon)$ only differs from its zero-temperature value over a small range of energy, of order $k_B T$, close to the Fermi surface (see Fig. 2.7).

Another important parameter is the so-called *density of states,* that is the number of levels per unit energy range. The density of states is proportional to the volume of the gas, but it is often convenient to work with a parameter which depends only on the density. In addition, for a degenerate Fermi system, it is generally sufficient to consider only the value of this parameter at the Fermi surface. We therefore introduce the quantity $(d\tau/d\epsilon)$, the density of states for unit volume of gas, which has the value at the Fermi surface of

$$\left(\frac{d\tau}{d\epsilon}\right)_0 = \frac{m}{\pi\hbar^2}\left(\frac{3N}{\pi V}\right)^{\frac{1}{3}} = \frac{m}{\pi^2\hbar^3}p_0. \tag{2.3}$$

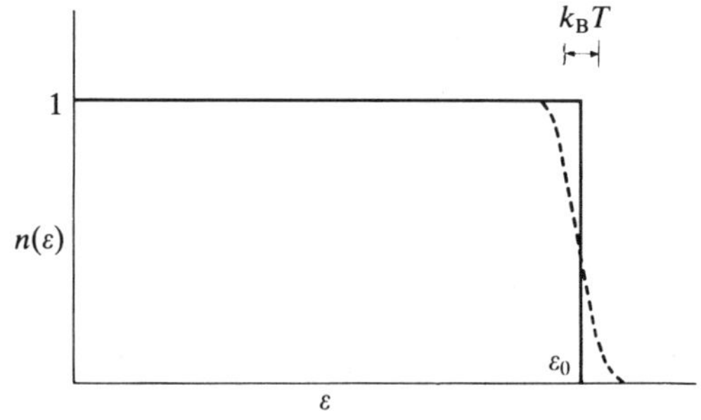

Fig. 2.7. Schematic diagram of the Fermi–Dirac distribution function. The full line shows the distribution for $T = 0$; the dotted line corresponds to a somewhat higher temperature T.

The specific heat per unit volume is obtained by differentiating the internal energy of the whole system, and is found to have the value

$$\mathscr{C} = \tfrac{1}{3}\pi^2 k_B^2 \left(\frac{d\tau}{d\epsilon}\right)_0 T. \tag{2.4}$$

We shall also find it useful to quote a result for the velocity of sound c, viz

$$c^2 = \frac{p_0^2}{3m^2}.$$

The magnetic susceptibility. In the presence of a magnetic field B, the energy of a level includes an additional term $-2\beta\sigma B$, where β is the nuclear magnetic moment, and σ the spin coordinate is either $\pm\frac{1}{2}$. Thus the effect of applying a magnetic field is to shift all the levels with spin $\frac{1}{2}$ relative to those with spin $-\frac{1}{2}$ by an amount $-2\beta B$. In equilibrium there is an excess number of atoms, Δ per unit volume, with spins parallel to the field where

$$\Delta = 2\beta B\left\{\frac{1}{2}\left(\frac{d\tau}{d\epsilon}\right)_0\right\}.$$

Hence there is a net magnetic moment $\beta\Delta$, and the susceptibility per unit volume is

$$\chi = \frac{\beta\Delta}{B} = \beta^2\left(\frac{d\tau}{d\epsilon}\right)_0,$$

which is independent of the temperature.

2.4 Liquid ^{3}He as a Fermi gas

Both liquid ^{3}He and a degenerate Fermi gas have a specific heat varying linearly with temperature, and a nuclear susceptibility which is independent of temperature. We now compare the numerical values of the specific heat C, the velocity of sound c, and the volume susceptibility χ of liquid ^{3}He under the saturated vapour pressure, with those for a perfect gas of the same density. The results of this comparison in the low-temperature region are shown in Table 2.1. We see that the properties of the liquid are of the same order of magnitude as those of the perfect gas, but that they differ by an appreciable factor.

The parallel between the behaviour of liquid ^{3}He and a Fermi gas is much strengthened when we consider transport coefficients such as the viscosity. According to the elementary kinetic theory of gases, the

Table 2.1

	Observed	Perfect gas	Ratio
C/RT	2.78	1.00	2.78
$c(\mathrm{m\,s^{-1}})$	183	95	1.92
χ(c.g.s.)	$3.3\times10^{38}\,\beta^2$	$3.61\times10^{37}\,\beta^2$	9.1

R is the gas constant, and β the magnetic moment of the ^{3}He nucleus

viscosity may be written

$$\eta = \tfrac{1}{3}\rho\tau v^2,$$

where τ (do not confuse with $d\tau$) is the mean time between collisions, and v the mean velocity of the atoms. We can show quite simply, for a gas obeying the Fermi statistics, that τ varies rapidly with temperature. When a collision occurs, two particles with energies ϵ_1 and ϵ_2 scatter into two different levels with energies ϵ_3 and ϵ_4 (such that $\epsilon_1+\epsilon_2=\epsilon_3+\epsilon_4$). According to the Fermi statistics, this process is only possible if the levels ϵ_3 and ϵ_4 are unoccupied before the collision. Now, the form of the distribution function is such that the only empty levels accessible to the scattering atoms are those in the range $\Delta\epsilon \simeq k_B T$ shown in Fig. 2.7. Therefore the probability of finding two empty levels is proportional to $(\Delta\epsilon)^2$, that is to $(k_B T)^2$. Hence we conclude that

$$\tau \propto 1/T^2.$$

Moreover, as the only atoms able to scatter are those close to the Fermi surface, their velocity will approximate to the value v_0 at the surface, and be independent of T. Hence we conclude that

$$\eta \propto \tau \propto 1/T^2.$$

The coefficients of *thermal conductivity* and *self diffusion*, K and D, are given by simple kinetic theory as

$$K = \tfrac{1}{3}\rho\tau v^2 C \quad \text{and} \quad D = \tfrac{1}{3}\tau v^2,$$

where C is the specific heat. Hence, by similar arguments to those used for the viscosity, we deduce that

$$K \propto 1/T, \qquad D \propto 1/T^2.$$

We cannot calculate the absolute values of these coefficients without knowing the collision cross-sections of the atoms, but the predicted dependences on temperature are similar to those observed in liquid ^{3}He. We conclude that the properties of the liquid are controlled to a large extent by the Fermi statistics. Thus a Fermi gas model of liquid ^{3}He gives a fair representation of its properties, but is by no means exact or complete.

2.5 The melting curve

The melting curve of ^{3}He has the same general form as that of ^{4}He, except that the melting pressure exhibits a shallow minimum of about 29.3 atm in the region of 0.32 K, as shown in Fig. 2.8. This minimum has interesting implications. Consider the Clausius Clapeyron relation

$$\left(\frac{\partial p}{\partial T}\right)_{\text{melt}} = \frac{S_1 - S_s}{V_1 - V_s},$$

where S and V are the molar entropies and volumes, and the subscripts 1 and s refer to the liquid and solid. The volume increase on melting is quite large, and remains so at the lowest temperatures. Hence the rise of the curve below 0.3 K ($\mathrm{d}p/\mathrm{d}T < 0$) implies that the entropy of the solid is greater than that of the liquid.

The above behaviour comes about because the atoms in the solid are more localized than in the liquid; their wave functions overlap less, and the exchange forces are correspondingly smaller. Hence the spins in the

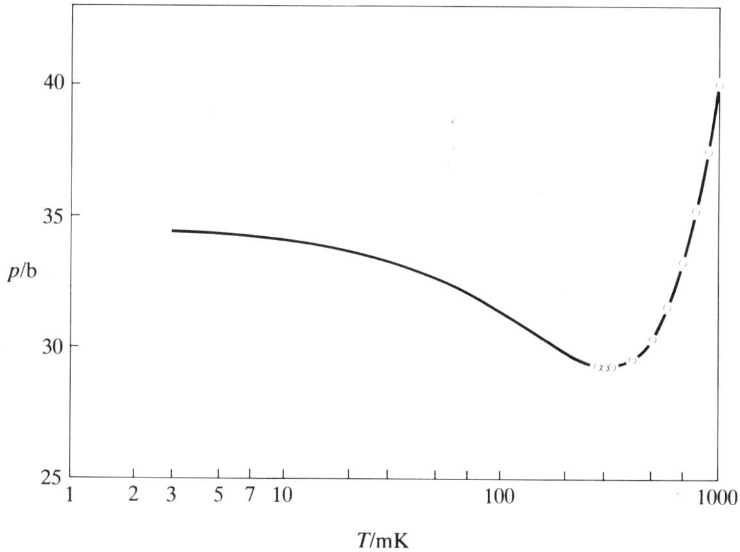

Fig. 2.8. The melting curve of ^{3}He. Experimental points represented by circles are due to Grilly [16]. Below the minimum the continuous line is derived from a polynomial fit to data by Greywall and Busch [17].

solid remain unaligned down to well below 0.01 K. Above this temperature, the entropy of the solid must therefore include a contribution of $R \ln 2$ from the spin disorder. Moreover, as the lattice entropy is very small below 1 K, the total entropy of the solid will be quite close to $R \ln 2$. In the liquid, the spin disorder is removed at much higher temperatures, and the entropy drops off rapidly. Hence the entropy of the liquid finally becomes less than that of the solid, and the melting curve has a negative slope. Eventually, of course, the slope of the curve must tend to zero at absolute zero in accord with the Third Law of Thermodynamics. At the minimum of the melting curve, the solid and liquid will have the same entropy, that is the entropy of the liquid should be approximately $R \ln 2$, as is observed.

As a point of experimental technique, we note that the minimum in the melting curve makes it somewhat difficult to measure the melting pressure at the lowest temperatures. In particular, the usual technique of measuring the pressure by observing when a capillary containing the liquid has been blocked by solidification is no longer useful. Suppose the coldest part of the capillary is at 0.1 K, where the melting pressure is about 31 atm. In order to leave the cryostat, the capillary must somewhere pass through a temperature of 0.32 K, where the melting pressure is only 29.3 atm. Hence, if the specimen is at a temperature below 0.32 K, the capillary will always block at about 29 atm. It is therefore necessary to work with a fixed volume of ^{3}He and measure the pressure either with strain gauges or with some form of bellows device. The melting curve of ^{3}He is now known so accurately that it can be used as a convenient means of precise thermometry [17].

The melting curve of ^{4}He also exhibits a minimum at about 0.8 K. However this is extremely shallow and difficult to observe compared to that in ^{3}He. It arises because the absence of transverse phonon modes in ^{4}HeII reduces the entropy of the liquid to a value below that of the solid, which does have such modes.

3
THE LANDAU THEORY OF A FERMI LIQUID

We now outline Landau's theory of a Fermi liquid, which commences with a gas-like model, and then takes account of the interactions between the atoms (Landau [18]).

3.1 The Landau model

The essence of Landau's model is illustrated schematically in Fig. 3.1. The left-hand side of the diagram shows the energy levels of a degenerate perfect gas governed by the Fermi statistics. The energies of the levels are uniquely determined by the properties of the individual atoms, and are calculated by solving the Schrödinger equation for one atom occupying the total volume. As the gas is assumed to be degenerate, the distribution function only varies in the region of the Fermi surface, as in Fig. 2.7. The momentum of an atom on the Fermi surface is, as previously noted,

$$p_0 = \pi\hbar\left(\frac{3N}{\pi V}\right)^{\frac{1}{3}}. \tag{3.1}$$

Landau next uses a perturbation method to introduce the interactions between the atoms. He assumes that these interactions can be gradually turned on, thus modifying the energy of each level. He also assumes that the wave number, or momentum, of each level is unchanged. In particular the momentum of a particle on the Fermi surface is still given by (3.1). When the full interaction has been introduced, each level has (in general) a different energy, as is indicated schematically on the right-hand side of Fig. 3.1. We now consider the properties of this new system.

The number of particles N in a volume V is given in terms of the distribution function n by

$$N/V = \int n \, \mathrm{d}\tau, \tag{3.2}$$

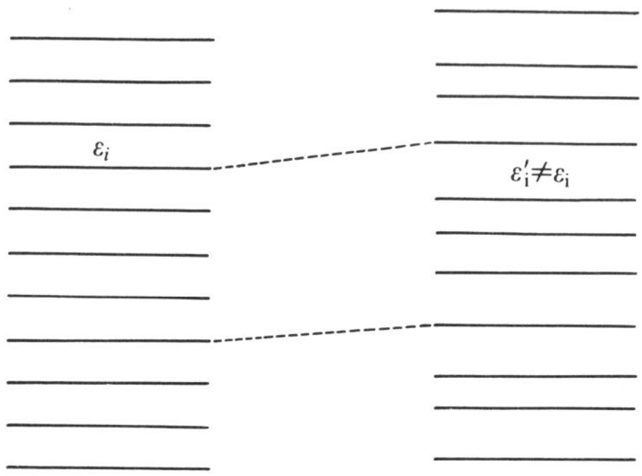

Fig. 3.1. Schematic diagram of the energy levels of a perfect gas and a Fermi liquid. Corresponding levels have the same wave number but different energies; see text.

the integration being taken over all momentum space so that

$$\mathrm{d}\tau = \frac{2\,\mathrm{d}p_x\,\mathrm{d}p_y\,\mathrm{d}p_z}{(2\pi\hbar)^3},$$

the factor 2 taking account of the spin degeneracy of the levels. In spite of the simplicity of Fig. 3.1, the system with interactions is much more complicated than the perfect gas. Because of the interactions the total energy cannot now be written as a simple sum

$$U/V = \sum_i n_i \epsilon_i,$$

where the n_i and ϵ_i are the occupation number and energy of the ith level. The energy of a level ϵ_i is no longer a unique function for an atom, but depends on the occupation numbers of all the levels. It is in fact no longer clear what we mean by the 'energy of an atom'. Landau therefore *defines* an energy by the relation

$$\delta U/V = \int \epsilon\,\delta n\,\mathrm{d}\tau, \tag{3.3}$$

where δU is the change in energy of the whole system when a small change δn is made in the distribution function. (In general this energy will depend on the nuclear spin, but for many purposes we can regard ^{3}He as an isotropic liquid, and then symmetry requirements imply that the energies do not depend on the spin parameter.)

The energy ϵ defined by eqn (3.3) is the energy of an atom interacting with all the other N particles *with a particular distribution.* Before turning on the interactions, the energy of the atom was related to its momentum and wave number by the usual energy spectrum

$$\epsilon = p^2/2m = \hbar^2 k^2/2m.$$

On introducing interactions, this simple relation characteristic of a real free particle is no longer valid, so we say that eqn (3.3) defines the energy of an *excitation* or *quasi-particle.* That is, we replace the system of real particles by a system of quasi-particles, which describe the atomic motion in a way which (as we show below) is tractable to calculation.

3.2 The distribution function and energy spectrum

The number $n(\epsilon)\,\mathrm{d}\epsilon$ of quasiparticles with energies in the range ϵ to $\epsilon + \mathrm{d}\epsilon$ is found in the normal way by maximizing the entropy subject to keeping the total energy and the total number of quasi-particles (equal to the number of atoms) constant. Hence

$$n(\epsilon) = \{\mathrm{e}^{(\epsilon-\mu)/k_\mathrm{B}T} + 1\}^{-1}, \tag{3.4}$$

where μ, the chemical potential, is determined by the condition

$$N/V = \int n\,\mathrm{d}\tau.$$

At first sight this distribution function is just the Fermi function for a perfect gas. However, relation (3.4) is much more complex, because ϵ is now a function of n, and the general form of this expression is rather unmanageable. However, the Landau theory is only concerned with the behaviour of the liquid at the lowest temperatures. In this case the distribution approximates to the usual Fermi step function, so it is approximately constant, and we may discuss the distribution in terms of its departure from the Fermi surface at temperature zero.

One effect of introducing the interactions is that (in general) the wave functions of the various levels will no longer be true eigenfunctions, and transitions will occur between the different levels. These transitions introduce an uncertainty $\delta\epsilon$ into the energy of a state, in accord with the uncertainty principle

$$\delta\epsilon \sim \hbar/\tau,$$

where τ is the lifetime of a state. This uncertainty places a restriction on the validity of the model. As the distribution function approximates to a Fermi function, the properties of the system are determined (like those of the gas) by the distribution of those quasi-particles lying in the energy range $\Delta\epsilon \sim k_B T$ close to the Fermi surface (Fig. 2.7). Clearly a discussion in terms of these energy levels is only meaningful if

$$\delta\epsilon < \Delta\epsilon. \tag{3.5}$$

It turns out that τ, which is equal to the time between collisions of the quasi-particles, varies as $1/T^2$ as for a perfect gas. Hence $\delta\epsilon$ is proportional to T^2, whereas $\Delta\epsilon$ is proportional to T. Therefore condition (3.5) can always be satisfied at a sufficiently low temperature.

We can now write down an expression for the energy spectrum, or dispersion relation, $\epsilon(p)$ of the quasi-particles near the Fermi surface. These particles lie in an energy range of order $k_B T \ll \mu_0$, so that at sufficiently low temperatures $\epsilon \simeq \mu_0$. It is then a good approximation to write

$$\epsilon = \mu_0 + \left(\frac{\partial\epsilon}{\partial p}\right)_0 (p - p_0), \tag{3.6}$$

where $(\partial\epsilon/\partial p)_0$ is evaluated at the Fermi surface. In the case of a perfect gas

$$\left(\frac{\partial\epsilon}{\partial p}\right)_0 = \frac{p_0}{m} = v_0,$$

where v_0 is the velocity of a particle at the Fermi surface. In the Landau theory $(\partial\epsilon/\partial p)_0$ has a different value, but by analogy we write

$$\left(\frac{\partial\epsilon}{\partial p}\right)_0 = \frac{p_0}{m^*} = v_0,$$

where m^* and v_0 are two new parameters. m^* is known as the effective mass, and the energy spectrum (3.6) now takes the form

$$\epsilon = \mu_0 + \frac{p_0}{m^*}(p - p_0). \tag{3.7}$$

According to Landau, the energy ϵ may be regarded as the Hamiltonian of a quasi-particle. In this case the relation $(\partial\epsilon/\partial p)_0 = v_0$ is the quantum version of Hamilton's equation $\partial H/\partial p = \dot{q}$, and v_0 is therefore taken to be the velocity of a quasi-particle on the Fermi surface.

Another important parameter is the density of states at the Fermi surface. This function is determined by the form of the energy spectrum at the Fermi surface, and its value is obtained by replacing m by m^* in the corresponding expression for a perfect gas, i.e.

$$\left(\frac{\mathrm{d}\tau}{\mathrm{d}\epsilon}\right)_0 = \frac{m^*}{\pi\hbar^2}\left(\frac{3N}{\pi V}\right)^{\frac{1}{3}}. \tag{3.8}$$

3.3 The interaction function

The energy ϵ of a level in the isotropic liquid is specified by the relation

$$\delta U/V = \int \epsilon \, \delta n \, \mathrm{d}\tau. \tag{3.3}$$

That is, the energy of a quasi-particle is the first functional derivative of the total energy with respect to the distribution function. We also require an expression for the change in energy of a particular level due to a redistribution of particles in the other levels. This is

$$\epsilon(\boldsymbol{p}) = \epsilon_0(\boldsymbol{p}) + \int f(\boldsymbol{p}, \boldsymbol{p}') \, \delta n' \, \mathrm{d}\tau',$$

where

$$f(\boldsymbol{p}, \boldsymbol{p}') = \frac{\partial^2 U}{\partial n(\boldsymbol{p}) \, \partial n'(\boldsymbol{p}')}.$$

In this relation $\epsilon_0(\boldsymbol{p})$ is the energy of the level with momentum $\boldsymbol{p}$, in an equilibrium distribution at temperature $T = 0$. The quantity $\epsilon(\boldsymbol{p})$ is the energy of the same level (with momentum $\boldsymbol{p}$) when the occupation of the other levels ($\boldsymbol{p}'$) differs from the zero-temperature distribution by an amount $\delta n'(\boldsymbol{p}')$. Thus the function $f(\boldsymbol{p}, \boldsymbol{p}')$ specifies how the energy of a level is modified by the occupation of the other levels. As we will be concerned only with particles close to the Fermi surface, $p \simeq p' \simeq p_0$; hence in the first approximation $f(\boldsymbol{p}, \boldsymbol{p}')$ is a function only of χ, the angle between $\boldsymbol{p}$ and $\boldsymbol{p}'$. That is,

$$f(\boldsymbol{p}, \boldsymbol{p}') = f(\chi).$$

In order to express the interaction function in its most convenient form, Landau introduces a new function F defined as

$$F(\chi) = \left(\frac{\mathrm{d}\tau}{\mathrm{d}\epsilon}\right)_0 f(\chi), \tag{3.9}$$

where $(\mathrm{d}\tau/\mathrm{d}\epsilon)_0$ is the density of states given by (3.8). He then expands $F(\chi)$ in Legendre polynomials so that

$$F = \sum_n F_n p_n(\cos\chi) = F_0 + F_1 \cos\chi + F_2\left(\frac{3\cos^2\chi - 1}{2}\right) + \cdots,$$

where the F_n are constants to be determined. This expansion is important because a comparison of theory with experiment suggests that the principal contributions to F come from the leading terms. Hence we are able to specify the complex interactions in the liquid by only the first two or three of the constants F_n, as we show in the next section.

To discuss the behaviour of liquid ^{3}He in a magnetic field, it is necessary to introduce the spin coordinate σ. We will not discuss this in detail, but note that Landau rewrites (3.3) as

$$\delta U/V = \tfrac{1}{2}\mathrm{Sp}_\sigma \int \epsilon\, \delta n\, \mathrm{d}\tau. \tag{3.10}$$

The terms ϵ and n are now matrices, and the symbol Sp_σ implies the summing of the diagonal elements of the matrix product within the integral. The factor $\frac{1}{2}$ appears because we have already taken account of the two spin states in our definition of $\mathrm{d}\tau$. Similarly we must rewrite equation (3.2) as

$$N/V = \tfrac{1}{2}\mathrm{Sp}_\sigma \int n\, \mathrm{d}\tau. \tag{3.11}$$

The spin coordinate affects the total energy of the system through its effect on the exchange energy, which is the additional term in the expression for the electrostatic interaction between the atoms which arises when we use symmetrical wave functions to describe a system of identical particles. The total wave function of liquid ^{3}He must be antisymmetric because it obeys Fermi statistics. Hence, if we turn over one spin and thus change the symmetry of the spin wave function, the symmetry of the spatial wave function must also change so that the total wave function remains antisymmetric. This change in the spatial function will alter the value of the exchange integral; hence the total energy is a function of the spin coordinates. Therefore it comes about that the state of lowest energy is generally one where the nuclear spins take up either a parallel or antiparallel alignment.

To introduce the exchange force into the Landau model, the interaction function $f(\chi)$ is extended to include a term depending on the spin coordinates $\boldsymbol{\sigma}$, $\boldsymbol{\sigma}'$, viz.

$$\mathscr{F}(\boldsymbol{p}, \boldsymbol{\sigma}; \boldsymbol{p}', \boldsymbol{\sigma}') = f(\boldsymbol{p}, \boldsymbol{p}') + \zeta(\boldsymbol{p}, \boldsymbol{p}')\boldsymbol{\sigma}\cdot\boldsymbol{\sigma}'.$$

We also introduce a function $G(\chi)$, analogous to $F(\chi)$ in eqn (3.9), so that

$$G(\chi) = \left(\frac{\mathrm{d}\tau}{\mathrm{d}\epsilon}\right)_0 \zeta(\chi) = G_0 + G_1 \cos\chi + \cdots$$

We are now able to specify the exchange interaction quite well by the values of the constants G_0 and G_1 in the two leading terms. (Note that accounts of Landau theory often use the notation F_n^s for F_n, and F_n^a for $\frac{1}{4}G_n$; see for example [19].)

3.4 The properties of liquid ^{3}He

We now relate the Landau model to the properties of the liquid. Straightforward calculations show that at sufficiently low temperatures the temperature dependences of the specific heat, susceptibility, and transport coefficients are the same as those of the perfect gas. However the absolute magnitudes are all different, being related in a rather complicated way to the parameters of the theory. The constants F_0, F_1, and G_0 are specified exactly by the relations

$$C = \frac{m^*}{m} C_g \qquad \frac{m^*}{m} = (1 + \tfrac{1}{3}F_1) \tag{3.12}$$

$$c^2 = \frac{p_0^2}{3m^2} \frac{1 + F_0}{1 + \frac{1}{3}F_1}. \tag{3.13}$$

$$\chi = \frac{m^*}{m} (1 + \tfrac{1}{4}G_0)^{-1} \chi_g, \tag{3.14}$$

where C_g and χ_g are the specific heat and susceptibility of a perfect gas of the same density. Values for these constants derived from the measured specific heat, velocity of sound, and susceptibility are given in Table 3.1. (The values given are taken from a discussion by Greywall [6])

The specific heat of the liquid is greater than that of the gas (equation (2.4)) because the interactions increase the density of states at the Fermi surfaces in the ratio m^*/m, and the specific heat is proportional to this quantity. So far as the susceptibility is concerned the effect of the interactions shows itself in two ways. Like the specific heat, it depends directly on the density of levels at the Fermi surface, and is therefore augmented by a factor m^*/m. There is also an additional effect arising

Table 3.1. Fermi liquid parameters for liquid ^{3}He at zero pressure

Molar specific heat		2.78 RT	
Velocity of sound		180 m s^{-1}	
m^*/m	2.76		
G_0	−2.80	F_0	9.15
G_1	−2.20	F_1	5.27

from the spin-dependent exchange forces, which give rise to the term $(1+\frac{1}{4}G_0)^{-1}$. Experimentally it is found that the susceptibility of liquid ^{3}He in the Fermi region is much higher than that of a perfect gas of the same density, so the constant G_0 is negative. Hence the exchange forces favour parallel alignment of the spins, in contrast to the Fermi statistics which tend to produce anti-parallel alignment. This effect is quite large, as the term $\frac{1}{4}G_0$ turns out to have a value of about $-0{\cdot}7$. Indeed the form of eqn (3.14) implies that if the exchange forces were about 50 per cent stronger ($\frac{1}{4}G_0 \leq -1$), the ground state of the liquid would be ferromagnetic.

The Landau theory also predicts the magnitude of the transport coefficients by setting up and solving a Boltzmann transport equation for the motion of the quasi-particles in steady but non-uniform conditions. The temperature dependence of the coefficients of viscosity η, thermal conductivity K, and self diffusion D, turn out to be similar to those of the perfect gas, in accord with the measured values. The absolute values of the transport coefficients are obtained after somewhat lengthy calculations, in terms of the parameters F_n and G_n. Substituting in the known values of F_0, F_1, G_0 and G_1, values are obtained for the coefficients of conductivity and diffusion which are in reasonable agreement with the measured values under pressures of 0·28 and 27 atm, particularly in view of the neglect of F_2, G_2, and higher terms. The theory also gives a fair account of the magnitude and pressure dependence of the viscosity. For some recent refinements of these calculations see reference [19].

We conclude that the Landau theory gives quite a satisfactory account of the properties of liquid ^{3}He. In particular it shows that the temperature dependence of the specific heat, susceptibility, and transport coefficients is a direct consequence of the Fermi distribution function. Thus liquid ^{3}He provides the most direct and striking example of the effect of the Fermi statistics.

3.5 Zero sound

We have now shown that the Landau theory gives a comprehensive account of the specific heat, velocity of sound, susceptibility, and transport properties. However, as F_0, F_1, and G_0 are used as independent adjustable parameters, we have not made too vigorous a verification of the theory. A crucial feature of the Landau theory is that it predicts an entirely new mode of sound propagation, the so-called zero sound (Landau [20]).

As already mentioned, the relaxation time characterizing the collisions between the quasi-particles varies as $1/T^2$. Hence, at sufficiently low temperatures, the time between collisions becomes greater than the period of any sound wave. At first sight it would appear that the propagation of sound is then impossible. The position would be partly analogous to attempting to propagate a sound wave in a rarefied gas with a mean free path greater than the wavelength of the sound. In fact a perfect Fermi gas cannot support a wave motion at very low temperatures. However, liquid ^{3}He does not behave as a perfect gas, because of the strong interactions between the atoms.

Landau's treatment starts from the Boltzmann equation for the distribution function of the quasi-particles, and derives the response of the distribution function to a periodic disturbance of angular frequency ω. In the limit $\omega\tau \ll 1$ the disturbance is ordinary hydrodynamical sound. However, other wave solutions can be obtained if $\omega\tau \gg 1$, the possibility of observing these solutions depending on the strength of the interactions, and Landau named these modes 'zero sound'. The theory predicts a longitudinal zero sound mode analogous to ordinary sound, a transverse mode, and collisionless spin waves. These have all been observed ([21–24]) but we concentrate here on longitudinal zero sound.

It turns out that two characteristics of the zero-sound mode are a different velocity and a different coefficient of absorption from those of ordinary sound. In general, the velocity will depend on the values of all the constants F_n; however, if we use the only values we know (F_0 and F_1), we find that the velocity in the zero-sound region should be greater by about 6 m s^{-1}. The absorption in the ordinary region arises principally from the viscosity and may be written as

$$\alpha = A\omega^2\eta,$$

where A is a constant, and η, the viscosity, varies as $1/T^2$. The behaviour in the zero-sound region is quite different, viz.

$$\alpha_0 = A_0 T^2,$$

where A_0 is a constant which is independent of the frequency.

Longitudinal zero sound was first detected by measuring the reflection coefficient of a sound wave at the boundary between a quartz crystal and the liquid. This coefficient is directly related to the acoustic impedance of the liquid, Z, which for an ordinary liquid has the value ρc, where ρ is the density, and c the velocity of sound. As the temperature fell from 0.1 to 0.08 K the value of Z/ρ for a 1 GHz sound wave increased by about 19 m s^{-1}. Also by estimating a value of τ from the viscosity it appears that at the transition $\omega\tau$ is of order unity as predicted [21, 10].

Figure 3.2 shows subsequent measurements of the velocity and absorption. The transition again occurs at the expected temperature, and the change in velocity is 6 m s^{-1}, the value we calculate by ignoring F_2 and higher terms. Note in particular that the absorption below the maximum

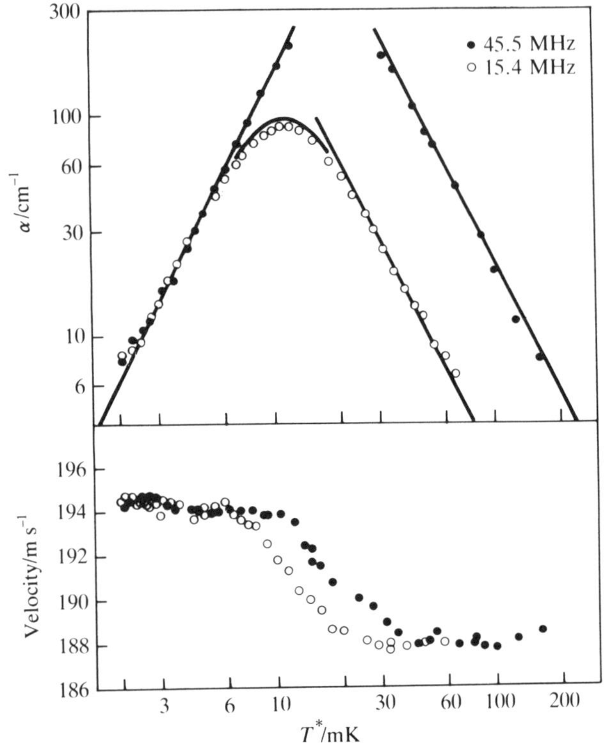

Fig. 3.2. The amplitude attenuation coefficient α and the velocity of sound and zero sound in liquid ^{3}He (Abel *et al.* [22]).

has the expected dependence on frequency and temperature. Thus the principal features of zero sound are correctly predicted but various points of interest remain. In particular the theory leads us to expect that the value of Z/ρ should be equal to or less than the velocity of zero sound c_0, and independent of the frequency, whereas experiments show that Z/ρ is appreciably greater than c_0, and may depend on the frequency of the sound. These points are further discussed in reference [25].

3.6 Developments of the Landau Theory

The original formulation of the Landau theory has been extended and improved in a number of ways (see for example Baym and Pethick [19]). Improvements in methods of calculation have given more exact magnitudes of the transport properties, and the theory has been extended to higher temperatures by including the effects of spin fluctuations or paramagnons. We saw in Section 3.4 that the exchange forces in liquid ^{3}He tend to align the ^{3}He nuclei in a ferromagnetic arrangement in opposition to the rather larger effect of the Fermi statistics, which tend towards an antiparallel alignment. In fact if the exchange forces were 50 per cent stronger the liquid would be a ferromagnetic. It is believed that fluctuations in small regions of the liquid will momentarily show ferromagnetic alignments and thus modify the entropy and other thermodynamic properties of the liquid. The effect of these fluctuations leads to the inclusion of the term $\Gamma T^3 \ln(T/\theta_c)$ in equation (2.1) for the specific heat, as discussed by Baym and Pethick [19]. Recent data support the validity of this term up to about 100 mK as shown in Fig. 2.2. The fluctuations may also scatter the quasi-particles in transport processes, and so modify the kinetic coefficients, see [19].

The Landau theory cannot make predictions about the numerical values of the Landau interaction parameters F_0, F_1, G_0, G_1, etc. since they are undetermined within that theory. It is attractive to suppose that the behaviour of liquid ^{3}He could be rigorously deduced from a basic minimum amount of information including the best available interaction

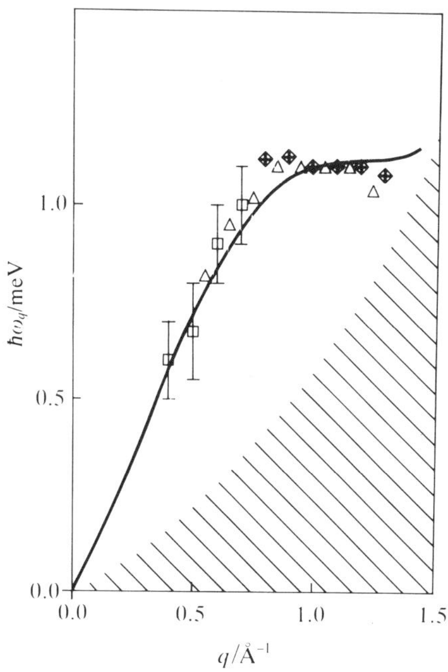

Fig. 3.3. Measurements of the dispersion curve of zero-sound excitations [32, 33] are shown as data points. The continuous line is from theory [34] and the shaded region indicates the quasi-particle continuum discussed in the text. After Aldrich and Pines [34].

potential between the atoms (Aziz *et al.* [26]), internuclear dipole–dipole forces, and the rules of quantum mechanics. An exact treatment is still remote but approximate methods are now able to make reasonable estimates of the principal Landau parameters (see for example [27–29]).

The elementary excitations in liquid ^{3}He have been studied by neutron scattering experiments. These were similar in principle to those on liquid ^{4}He described in Section 6.3, but the experimental difficulties were considerably greater because the ^{3}He nuclei have large cross-sections for capture. The method consists of directing a beam of monochromatic neutrons into the liquid and studying the angular dependence of the energy spectrum of the scattered neutrons. With an ordinary liquid the neutrons scattered through a given angle exhibit a broad spectrum of energies because they may follow a large number of different collision sequences. In liquid ^{4}He on the other hand the spectrum is quite sharp, with only a narrow range of energies at each angle because ^{4}HeII has rather well-defined excitations as discussed later. Normal liquid ^{3}He exhibits both sharp and broad features in the spectra. The continuous line in Fig. 3.3 shows the dispersion curve associated with the zero-sound modes, and the shaded area the so-called 'quasi-particle continuum'. The spectra in the continuum region are broad corresponding to the range of energy levels near the Fermi surface into which quasi-particles can be scattered by the incident neutrons. The detailed interpretation of this continuum is somewhat problematic although Aldrich *et al.* [30] have calculated mean energies in agreement with values observed by Hilton *et al.* [31].

4
LIQUID HELIUM II

Like liquid ^{3}He at higher temperatures, liquid ^{4}He near its boiling point may be regarded as behaving rather like a dense gas of classical particles. However, at a temperature of 2.17 K, there is an abrupt change in all the properties of the liquid. This change is known as the lambda transition, after the characteristic form of the specific heat curve (Fig. 1.3), which rises to a very high value at the lambda temperature of 2.17 K. The liquid behaves so differently above and below this temperature that we refer to it in the two regions as *helium I* and *helium II* respectively. The difference is apparent visually if we observe a dewar vessel of liquid ^{4}He which is cooled through the lambda point by boiling under reduced pressure. In the helium I region, the liquid is greatly agitated by bubbles of vapour which form throughout the liquid. Yet, immediately the lambda point is passed, the liquid becomes quite calm and still. Further investigation shows that heat transport takes place so readily within the fluid that it is almost impossible to set up appreciable temperature gradients; hence all the evaporation takes place from the free surface, without bubbling. Many of the key papers in the history of the understanding of liquid ^{4}He have been conveniently collected by Galasiewicz [35].

4.1 Viscosity and superfluidity

Measurements of the flow of helium II through very fine capillaries [36] indicate a vanishingly small viscosity, at least 10^{11} times smaller than that of helium I (Fig. 4.1(a)). This is the so-called *superfluidity* of liquid helium II, perhaps its most characteristic feature. For not too great pressures, the velocity of flow through these fine capillaries appears independent of the pressure head, and is actually *greater* in tubes of *smaller* diameter.

The viscosity has also been measured with rotation viscometers. The results are quite dissimilar from those described above; in particular there appears to be no difficulty in observing a normal coefficient of viscosity. Figure 4.1(b) shows values of the viscosity obtained in this way; the magnitude is of the same order as that observed in helium I, although the dependence on temperature is different. Figure 4.1(c) shows values of the

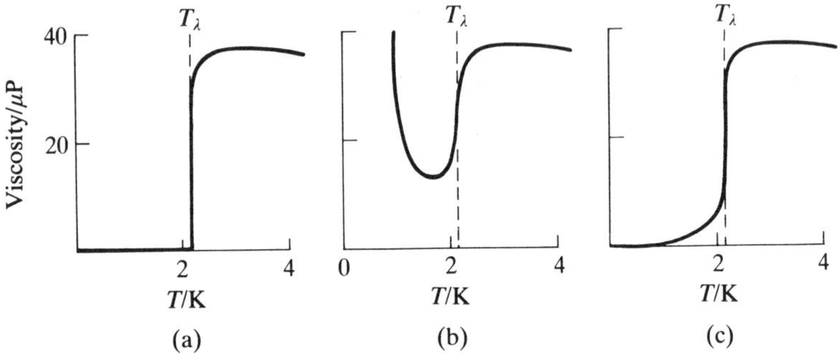

Fig. 4.1. The apparent viscosity of liquid ^{4}He measured (a) by a Poiseuille flow method, (b) in a rotation viscometer, (c) in an oscillating disc viscometer.

viscosity derived in the usual way from the observed damping on a disc oscillating in the liquid. For conventional liquids, including liquid helium I, we expect to obtain the same values for the viscosity in both a rotation and oscillation viscometer, but Fig. 4.1 shows that this is not so for helium II.

A characteristic feature of the behaviour of helium II is that the damping on an oscillating disc is a function of the amplitude of the oscillations. The damping is independent of amplitude over a large range of amplitude, and then rapidly increases, rather as if some form of turbulence had set in quite abruptly. We return to this point in Section 4.3.

Associated with superfluidity is the behaviour exhibited in the remarkable beaker experiment illustrated in Fig. 4.2. A small beaker or test tube

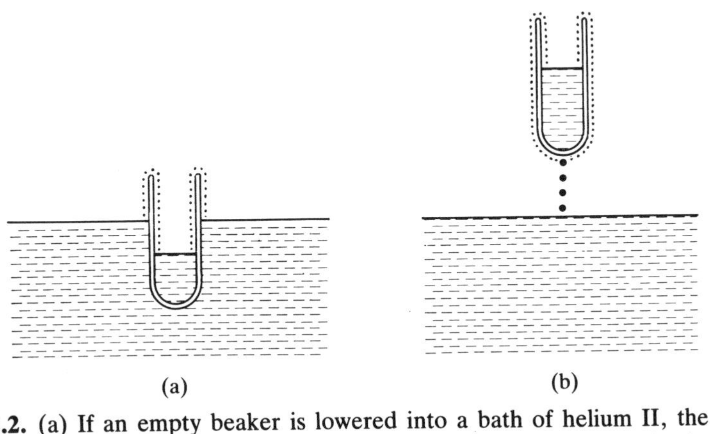

Fig. 4.2. (a) If an empty beaker is lowered into a bath of helium II, the liquid flows over the surface of the beaker until the levels are equalized. (b) If the beaker is then lifted above the bath, the helium flows out over the rim, drops off the bottom of the beaker, and returns to the bath (after Daunt and Mendelssohn [37]).

containing helium II is suspended above a bath of helium II at the same temperature. It is observed that the liquid in the beaker falls to a lower level by flowing up and over the edge of the beaker, then down to the bottom of the tube, where it forms drops which fall off into the bath.

4.2 The thermomechanical effect

Another property is illustrated by the simple experiment of Fig. 4.3. A and B are two vessels joined by a fine capillary. Initially both vessels contain helium II, at the same temperature, and standing at the same level. If an excess pressure Δp is applied over the helium in vessel A, liquid flows through the capillary into the vessel B, where the level will stand above that in A. In addition, however, the temperature of A rises a little, and that of B falls. Moreover, on releasing the excess pressure, the system returns to its original condition. This result implies, unequivocally, that the liquid which passes through the capillary is different from the bulk liquid in the vessels A and B. In fact, the properties of helium II are often described in terms of a 'two-fluid model', the liquid being regarded as the sum of two 'fluids', one of which flows through the capillary while the other does not. In another version of this experiment we start with the helium in vessels A and B, at the same level and the same temperature, and then raise the temperature of one vessel by a small amount ΔT. We then observe a change in level corresponding to a pressure change Δp, the relative magnitudes of Δp and ΔT being similar to those observed previously. This phenomenon is known as the thermomechanical effect, but is also sometimes referred to as the fountain effect, because so much heat was supplied in one of the earliest experiments, that the liquid was thrown up in a jet.

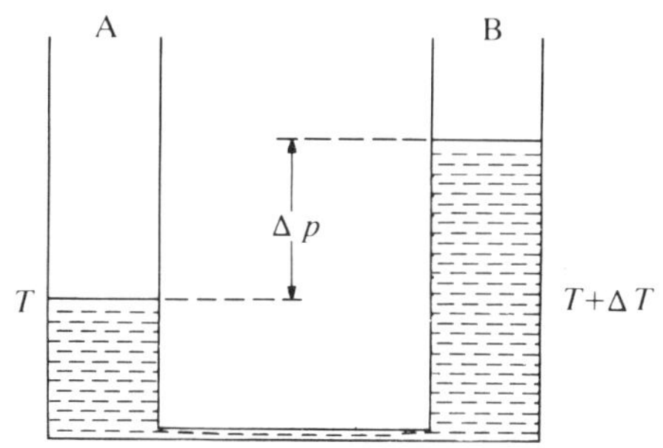

Fig. 4.3. The vessels A and B are joined by a fine capillary. A displacement of liquid helium II from A to B creates a temperature difference between the two vessels. Alternatively, a temperature difference ΔT results in a hydrostatic pressure difference Δp.

4.3 Heat transport

The behaviour of the flow of heat in helium II is also remarkable. As the transfer of heat is very high, many measurements have been made in long narrow glass capillaries, any heat flow in the glass being negligible compared with that in the helium. Some early results are shown in Fig. 4.4. The heat flow is very high, passes through a maximum at about 1.8 K, and appears to be proportional to $(\mathrm{d}T/\mathrm{d}x)^{\frac{1}{3}}$, the cube root of the temperature gradient. Clearly we are dealing with a very complicated phenomenon. However, if measurements are made with not too large heat flows, in very narrow channels, then the flow is proportional to the temperature gradient. For example Fig. 4.5 shows some results where the heat flow is proportional to the temperature difference up to a certain value, but then increases less rapidly. Even in this linear region, however, the heat flow is far from normal. It is extremely high, and not proportional to the area of the slit.

In order to discuss the complex behaviour of helium II in detail it is convenient to consider various patterns of behaviour separately. We have just seen that at sufficiently low velocities the damping on an oscillating

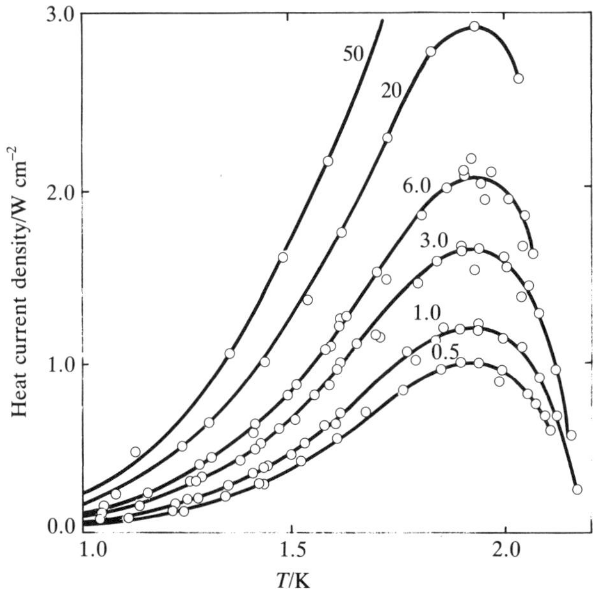

Fig. 4.4. The heat current density produced by different temperature gradients in capillaries of the order of 1 mm diameter. The numbers on each curve give the temperature gradient in $10^{-3}\,\mathrm{K\,cm^{-1}}$ (after Keesom *et al.* [38]).

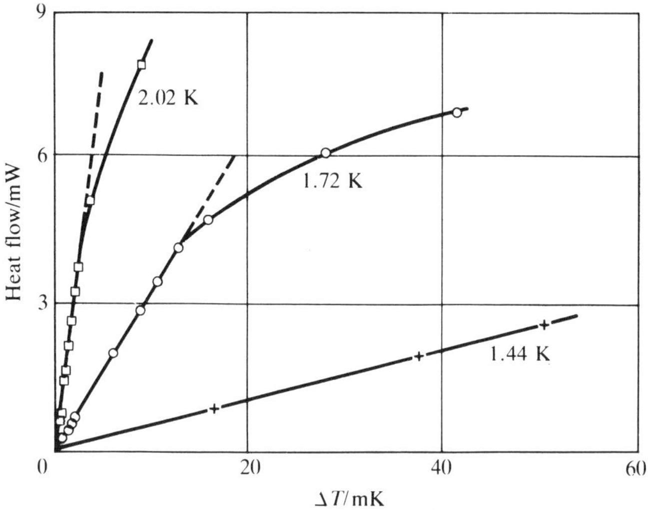

Fig. 4.5. The heat flow through liquid helium in a slit of width 2.4 μm as a function of the temperature difference ΔT across the slit. The relation is linear for small heat flows at not too high temperatures (after Winkel *et al.* [39]).

disc is independent of velocity, and the heat flow is proportional to the applied temperature difference. In fact, it is found that in a wide range of experiments, similar linear relations are valid, provided that the fluxes of heat or momentum are not too large. We shall describe this condition of the liquid as the *linear region*. This region is not to be regarded as one in which the additional effects observed at higher fluxes are small, but rather as one in which they are entirely absent. This point is illustrated by measurements on the flow of heat in a capillary 10 cm long and about 0.1 mm in diameter (Fig. 4.6). The thermal resistance of the liquid is initially independent of the heat flux, and then rises quite suddenly. As a convenient terminology we say that the region above the linear region is characterized by some form of *turbulence* in the liquid, and that this sets in quite abruptly at some *critical velocity*, the nature of which is still to be specified.

In the following chapters we discuss the large number of experiments which have been made in the linear region, and their interpretation in terms of a 'two-fluid' model. The behaviour of helium II in the turbulent region turns out to be ultimately related to the fact that many of its kinetic properties are modified when the liquid is set in rotation. This we discuss in Chapters 12 to 14.

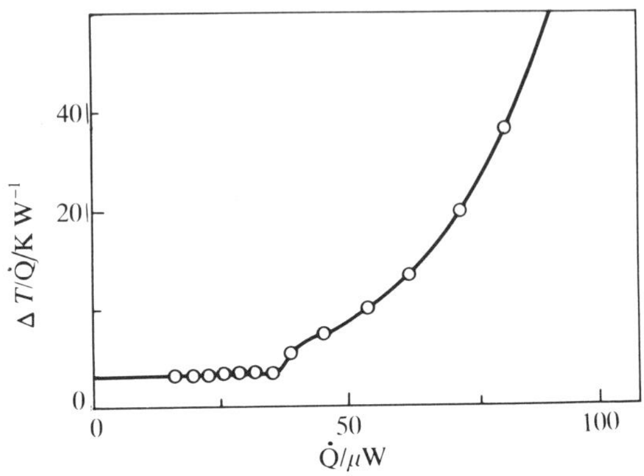

Fig. 4.6. The thermal resistance $\Delta T/\dot{Q}$ of a capillary filled with liquid helium II (107.6 μm diameter, 10 cm long), as a function of heat current $\dot{Q}$ (Brewer and Edwards [40]).

4.4 Second sound

One of the most unusual features of helium II is that variations of temperature propagate through the liquid not according to the usual Fourier equation, but as a true wave motion whose velocity is independent of the frequency. These temperature waves are entirely analogous to ordinary sound waves, save that the thermodynamic variable is temperature and not pressure. Thus we can excite temperature waves with a heater in a resonance tube, and detect standing waves using a thermometer as a detector, as shown in Fig. 4.7. Alternatively this wave-like

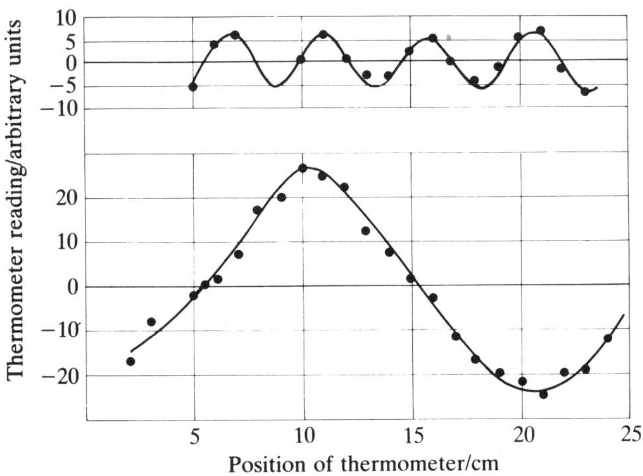

Fig. 4.7. Typical standing wave patterns of second sound (Peshkov [41]).

transmission of heat enables us to transmit sharp pulses of temperature as compact pulses through the liquid. This very unusual type of heat propagation is known as second sound. Other modes, known as third, fourth, and fifth sound, can occur in restricted environments such as surfaces or narrow capillaries. All these modes have been reviewed by Rudnick [42].

4.5 The momentum of heat flow

The flow of heat in helium II is accompanied by a flow of momentum. This momentum was first observed by Kapitza, who showed that a heat current impinging on a vane mounted on a torsion wire gave rise to a pressure which deflected the vane. A later experimental arrangement is shown schematically in Fig. 4.8. A flat heater wound from thin wire is

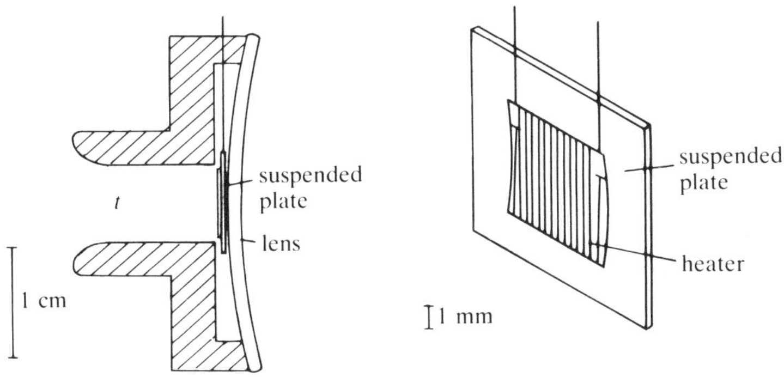

Fig. 4.8. Apparatus to observe the momentum associated with heat flow in helium II (Hall [43]).

mounted on one side of a thin glass plate suspended close to a convex lens, the whole assembly being immersed in a bath of helium. The low conductivity of the glass plate ensures that virtually all the heat flow from the wires takes place uni-directionally down the tube *t*; the plate therefore tends to move to the right. The displacement, and hence the effective force, is found by observing Newton's rings formed between the glass plate and the lens.

5
THE TWO-FLUID PROPERTIES OF HELIUM II

Until Chapter 13, we shall confine our attention to experiments in which the heat and mass currents are not too large. That is, we shall deal with helium II in the well-defined *linear region* mentioned in Section 4.3. It turns out that, in this region, many of the unusual properties of the liquid may be correlated with each other, and with the thermodynamic functions, by a mathematical picture known as the 'two-fluid model'.

5.1 The two-fluid model

The two-fluid model postulates that liquid helium II behaves as if it were a mixture of two fluids freely intermingling with each other without any viscous interaction. These two fluids are termed the *normal fluid* and the *superfluid,* and have densities ρ_n and ρ_s such that

$$\rho_n + \rho_s = \rho, \tag{5.1}$$

where ρ is the ordinary density of liquid helium. The normal density ρ_n is a function of temperature, and *increases from zero at absolute zero, to the value ρ at the lambda point.* Conversely, the superfluid density is zero at the lambda point and increases to the value ρ at temperature zero.

In addition, the model postulates that the superfluid carries zero entropy, and experiences no resistance whatever to its flow, that is, it exhibits neither viscosity nor turbulence. This condition is specified by stipulating that the viscosity of the superfluid is zero, and that its velocity $\boldsymbol{v}_s$ satisfies the relation

$$\text{curl}\, \boldsymbol{v}_s = 0.$$

On the other hand the normal fluid has a viscosity, the so-called *normal viscosity* η_n, and an entropy S_n equal to the entropy of the liquid helium. Thus the essential parameters of the model are shown in the following table.

Normal fluid	ρ_n	$\eta = \eta_n$	$S_n = S_{He}$
Superfluid	ρ_s	$\eta = 0$	$S_s = 0.$

The parameters ρ_n and ρ_s, and η_n, are functions of temperature but, at a given temperature, are taken as constants of the model. Their values as functions of temperature are deduced from some of the experiments described in the following sections. After these parameters have been determined, the model then correlates a wide variety of experimental results, without introducing any additional terms or adjustable constants. (The total density ρ also varies slightly with temperature, but the variation is very small, and to a good approximation may be ignored.)

We must stress that our use of the model only implies that helium II behaves *as if it were* a mixture of two fluids. It does *not* imply that helium II is a mixture of two real physical fluids. The liquid is an assembly of ^{4}He atoms which are all identical. We discuss how the mathematical model arises from physical principles in Chapters 6 and 7.

5.2 The equations of motion

We now derive equations of motion for the two fluids of the model, as given by Landau [44]. Let the symbol $\boldsymbol{j}$ denote the momentum of unit volume of liquid helium, and $\boldsymbol{v}_n$ and $\boldsymbol{v}_s$ the velocities of the two fluids, then

$$\boldsymbol{j} = \rho_n \boldsymbol{v}_n + \rho_s \boldsymbol{v}_s. \tag{5.2}$$

The flow $\boldsymbol{j}$ is also related to the density of the helium by the equation of continuity

$$\operatorname{div} \boldsymbol{j} = -\frac{\partial \rho}{\partial t}. \tag{5.3}$$

If the velocities of the two fluids are small as is often the case, we may ignore viscous effects and terms quadratic in the velocities. Euler's standard equation of motion then reduces to

$$\frac{\partial \boldsymbol{j}}{\partial t} = -\operatorname{grad} p, \tag{5.4}$$

where p is the pressure.

In the approximation which ignores the viscosity, the motions of the two fluids are reversible so that entropy is conserved. Thus, bearing in mind that the entropy of the liquid is associated only with the normal fluid, we have

$$\frac{\partial(\rho S)}{\partial t} = -\operatorname{div}(\rho S \boldsymbol{v}_n), \tag{5.5}$$

where S is the entropy per gram of liquid, and ρS the entropy per unit volume.

Finally we need an equation for the forces acting on the superfluid alone, and therefore write down the usual differential expression for the change in the internal energy of a mass of liquid helium when a change is made in the entropy, volume, and mass,

$$\mathrm{d}U = T\,\mathrm{d}S - p\,\mathrm{d}V + G\,\mathrm{d}M.$$

In this expression G is the Gibbs free energy of one gramme of liquid, i.e. the chemical potential, and $\mathrm{d}M$ is a change in the mass of the assembly. Let us now increase the mass of the assembly at constant volume by introducing particles which contribute only to the ground state, that is to the superfluid. As $\mathrm{d}S = \mathrm{d}V = 0$, it follows that the increase in internal energy is $G\,\mathrm{d}M$. Hence the work done in transferring a mass ΔM of superfluid from a point 1 in the liquid to an adjacent point 2 a distance $\mathrm{d}x$ away, a process which involves no change of entropy, is

$$\Delta W = \Delta U = (G_2 - G_1)\Delta M = \operatorname{grad} G \cdot \mathrm{d}x \cdot \Delta M.$$

It follows that the equation of motion of the superfluid may be written

$$\frac{\mathrm{d}\boldsymbol{v}_\mathrm{s}}{\mathrm{d}t} = -\operatorname{grad} G.$$

Expressing grad G in terms of standard thermodynamic identities, and again ignoring terms quadratic in the velocities, the last equation becomes

$$\frac{\partial \boldsymbol{v}_\mathrm{s}}{\partial t} = S \operatorname{grad} T - \frac{1}{\rho}\operatorname{grad} p. \tag{5.6}$$

By combining eqns (5.6) and (5.4), making use of eqn (5.2), and treating ρ_n and ρ_s as constants,

$$\frac{\partial \boldsymbol{v}_\mathrm{n}}{\partial t} = -\frac{\rho_\mathrm{s}}{\rho_\mathrm{n}} S \operatorname{grad} T - \frac{1}{\rho}\operatorname{grad} p. \tag{5.7}$$

Equations (5.1)–(5.7) are the basic equations of the two-fluid model in the linear approximation. We now apply them to the properties of helium II.

5.3 Second sound

Equations (5.3) and (5.4) lead readily to the relation

$$\frac{\partial^2 p}{\partial x^2} = \frac{1}{c^2}\frac{\partial^2 p}{\partial t^2},$$

where

$$c^2 = \left(\frac{\partial p}{\partial \rho}\right)_S,$$

the usual equation for the propagation of density or pressure changes in a liquid. However, there is the additional possibility, with the two-fluid model, of a fluctuation in the relative densities of the normal and superfluids, while the total density $(\rho_s + \rho_n)$ remains constant. As a change in the relative densities of normal and superfluids is equivalent to a change in temperature, an oscillation of the two fluids relative to each other corresponds to an entirely new phenomenon, a true oscillation of temperature. Remembering that the thermal expansion coefficient of helium is small, we can solve the five equations of the model and obtain the ordinary sound equation as before, plus a second relation

$$\frac{\partial^2 S}{\partial x^2} = \frac{\rho_n}{\rho_s} \frac{C}{TS^2} \frac{\partial^2 S}{\partial t^2},$$

C being the specific heat of unit mass of helium. This is a wave equation for the propagation of variations of entropy, or, what amounts to the same thing, variations of temperature. Thus the model predicts the existence of harmonic temperature waves with a velocity

$$c_2 = \left(\frac{\rho_s}{\rho_n} \frac{TS^2}{C}\right)^{\frac{1}{2}}.$$

The velocity has been measured both by standing wave and pulse techniques, and its behaviour is shown in Fig. 5.1. From these values we may deduce values of ρ_n which are in good agreement with those

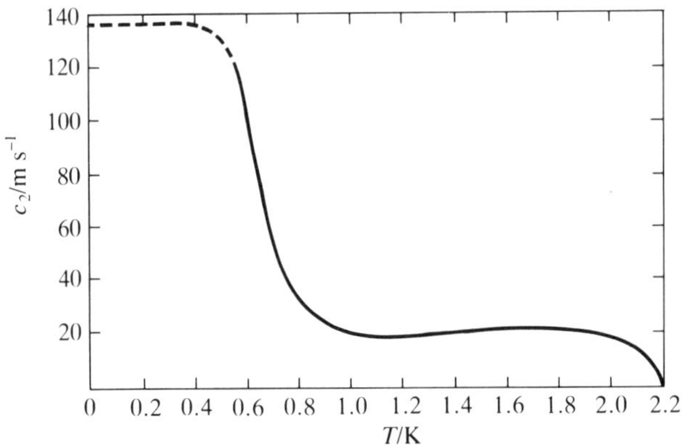

Fig. 5.1. The velocity of second sound (after Peshkov [41] and Khalatnikov [45]). The dashed section at low temperatures represents a region where it is generally difficult to observe second sound (cf. Section 7.5). Theory suggests that the velocity should approach a steady value equal to the velocity of ordinary sound divided by $\sqrt{3}$, or about 135 m s^{-1} as shown.

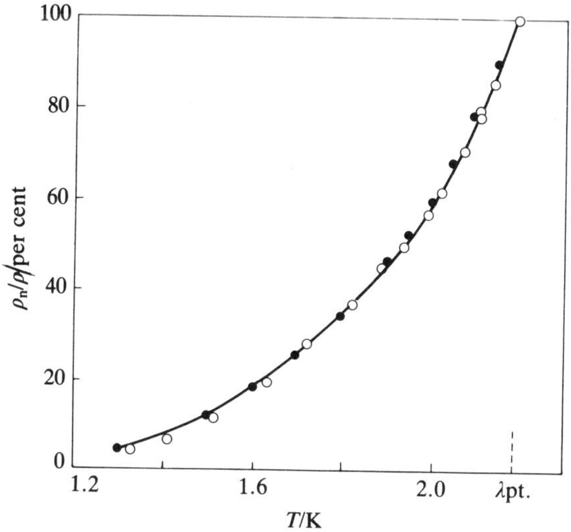

Fig. 5.2. The density of the normal fluid ρ_n as a function of temperature: ○ derived from oscllating disc experiments, ● from the velocity of second sound (Andronikashvili [46]).

obtained from other methods described below, see Fig. 5.2. In fact, measurements of the velocity give the most accurate method of determining ρ_n.

Historically, it should be mentioned that the existence of temperature waves was first predicted by Tisza on the basis of a different and less satisfactory model. However, at the time of writing his first paper, Landau was apparently unaware of this work, and seems to have thought that the velocity c_2 would be associated with sound of the usual type but with another velocity—hence he gave it the name 'second sound'. It was only after a failure to excite such waves with a piezo-electric crystal, that it was realized that a much more effective technique would be to generate a periodically varying temperature.

5.4 The viscosity

We now use the model to explain the very different values of the viscosity given by three standard methods of measurement (Fig. 4.1). In a Poiseuille flow type of measurement, the motion of the normal fluid is restricted by its viscosity, but the superfluid moves readily with zero viscosity. Hence we observe a flow without friction. In a rotation viscometer the position is rather different. The superfluid has no viscosity and transmits no torque between the cylinders, hence the viscometer measures just η_n, the viscosity of the normal fluid (Figs. 4.1 and 8.1). We have now obtained all the parameters of the two-fluid model, which is thus completely specified. We next consider how the model accounts for a variety of other phenomena.

To account for the form of the damping on an oscillating disc, we recall that in an oscillating disc viscometer, the damping depends on the product of the density and the viscosity. For helium II the appropriate density is ρ_n, so the *apparent* viscosity decreases rapidly with falling temperature (Fig. 4.1(c)). Substituting numerical values for ρ_n and recalculating the viscosity, we obtain figures for η_n in reasonable agreement with those given by the rotation viscometer. (Later and more precise measurements of the viscosity obtained with an oscillating disc are given in reference [47].)

We note also that the beaker experiment illustrated in Fig. 4.2 is readily explained in terms of the superfluidity of the helium. The exposed surfaces of the beaker are in equilibrium with the saturated vapour of the liquid in the bath, and will therefore be covered with a film of adsorbed helium, which is of the order of 20 nm thick. This film forms a continuous link between the two levels of helium in Fig. 4.2. Moreover the helium II forming the film can flow in narrow channels as a superfluid without any friction. Hence the film acts as a form of syphon between the helium in the beaker and the bath, so the two levels tend to equalize. Thus the motion of helium provides one more example of a liquid, albeit a quantum liquid, finding its own level! We should perhaps point out that there is no particular mystery as to the nature of the driving force for the motion. It is the difference in gravitational potential which leads to a difference in the free energy of the two liquids. As the levels are linked by the film they tend to adjust themselves until their free energies are equal.

5.5 Andronikashvili's experiment

This experiment gives the most direct method of measuring the density of the normal fluid ρ_n. It uses an apparatus rather similar to an oscillating disc viscometer except that the oscillating disc is replaced by a cylindrical vessel, in which is rigidly fixed a pile of fifty thin metal discs, as shown schematically in Fig. 5.3. This whole unit is mounted on a fine torsion suspension so that it can oscillate in liquid helium with a period of about 30 s. According to the two-fluid model, the superfluid takes no part in this motion, but normal fluid is dragged round by the discs as in the oscillation viscometer.

The penetration depth for a viscous wave, of angular frequency ω, in a liquid of density ρ_n and viscosity η_n is of the order of $(2\eta_n/\rho_n\omega)^{\frac{1}{2}}$, and at all temperatures below the lambda point this distance is appreciably

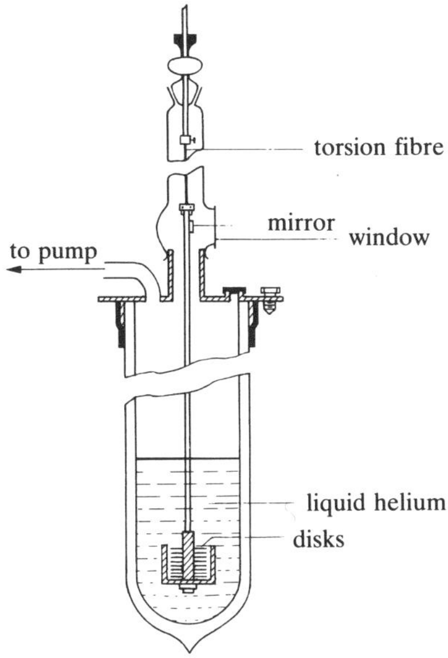

Fig. 5.3. Schematic diagram to illustrate Andronikashvili's experiment. Note that the discs are not drawn to scale; see text (Andronikashvili [46]).

greater than the spacing between the discs. Hence *all* the normal fluid between the discs is dragged round with them. Since the mass of this normal fluid varies with temperature, the apparent moment of inertia, and therefore the period of the system, also varies with temperature. Thus a measurement of the period, with and without helium present, offers a direct method of determining ρ_n.

The results obtained by Andronikashvili are shown in Fig. 5.2. The values are in good agreement with those derived quite independently from the velocity of second sound. The accuracy of the method falls off considerably at low temperatures, as the normal density becomes very small, so that the helium contributes very little to the moment of inertia of the oscillating system.

5.6 The thermomechanical effect

Let us consider the arrangement shown schematically in Fig. 4.3. Two baths of liquid helium II, A and B, at slightly different temperatures, T and $T + \Delta T$, are connected by a fine capillary. The capillary will greatly inhibit any flow of the normal fluid; on the other hand the superfluid, having zero viscosity, may pass freely. A temperature difference ΔT will initially set up a finite value of grad G along the tube, and this will produce an acceleration of superfluid from A to B according to eqn (5.6). Eventually a steady state is reached in which the superfluid moves through the tube at constant speed (the value of which will depend on the rate of the counterflow of the normal fluid as described in the next section). In this steady state with zero acceleration eqn (5.6) gives the relationship

$$\frac{\partial \boldsymbol{v}_s}{\partial t} = S \operatorname{grad} T - \frac{1}{\rho} \operatorname{grad} p = 0,$$

which leads to

$$\frac{\Delta p}{\Delta T} = \rho S, \tag{5.8}$$

a result first given by H. London. The effect is very large; for example at 1.5 K a temperature difference of only one millikelvin produces a pressure head of about 2 cm of liquid helium.

The fountain effect implies that if we let helium II run out of a container through a fine capillary, the emerging liquid will all be superfluid at temperature zero. In fact the outflowing liquid is never at absolute zero, both because of heat leaks and because there is always some flow of normal fluid in a capillary of finite diameter. However, a very considerable cooling of the emerging liquid is observed, while the temperature of the remaining liquid rises. Note that this method of producing very low temperatures is not of much practical use, as below 1 K the helium has very little entropy, and therefore very little capacity to cool other substances.

5.7 Heat transport

We consider the two reservoirs of helium at temperatures T and $T + \Delta T$ which were shown in Fig. 4.3. These two reservoirs are connected by a narrow capillary across which is developed a fountain pressure $\Delta p = \rho S \, \Delta T$. When discussing the fountain effect, we assumed the capillary to be infinitely narrow, but in fact it will allow some flow of the normal fluid under the action of the pressure Δp. Hence, even in equilibrium conditions, there will be a steady flow of normal fluid from the hot to the cold reservoir, and a return flow of super-fluid to conserve mass. As the normal fluid carries all the entropy, and the superfluid none, the net effect of this flow is to transfer entropy from the hot to the cold reservoir. This entropy transfer manifests itself as a large flux of heat.

The rates of flow of the two fluids in the capillary are limited only by the viscosity of the normal fluid. The volume rate of flow of this component is given by Poiseuille's formula to be

$$\dot{V}_n = \frac{\beta}{\eta_n}\frac{\Delta p}{\Delta x}, \tag{5.9}$$

where Δx is the length of the channel, and β a constant depending on its geometry. Since the superfluid carries no entropy, the rate of entropy transport between the reservoirs is $\dot{V}_n\rho S$. Hence, assuming as usual that the motion of the two fluids is reversible, the heat flow $\dot{Q} = T(\dot{V}_n\rho S)$. Substituting (5.9) into this expression, and using the London relation (5.8) for the fountain pressure, we finally obtain

$$\dot{Q} = \frac{\beta T(\rho S)^2}{\eta_n}\frac{\Delta T}{\Delta x}. \tag{5.10}$$

The verification of equation (5.10) is not entirely simple, as temperature differences of only a few millikelvins produce large fluxes of heat. Also the constant β contains the fourth power of the radius of the capillary (or the third power of the width of a rectangular slit). However, careful measurements have now shown that the relation is correct to within at least a few parts per cent. That is, the viscosity η_n can be correctly deduced from measurements of the heat flow. Note that the relation involves no adjustable constants.

Finally we note that, in the linear region, the heat flow for a given temperature gradient steadily increases with rising temperature right up to the lambda point in agreement with eqn (5.10). The maximum in the apparent conductivity shown in Fig. 4.4 is only characteristic of higher heat flows in the nonlinear region which we consider in Chapter 13.

5.8 The momentum of heat flow

As we have seen in the previous section, the flow of heat in helium II is accompanied by motion of the two fluids. Therefore, as the two fluids have momentum, the two-fluid model predicts that a flux of heat is accompanied by fluxes of momentum. This is the momentum already referred to in Section 4.5. The momentum flux associated with each fluid has the form $\rho \boldsymbol{v} \cdot \boldsymbol{v}$; thus in a unidirectional flow of heat there is a reaction pressure

$$p = \rho_n v_n^2 + \rho_s v_s^2$$

on the source due to the inertia of the fluids. The heat flow per unit area $W = \rho S T v_n$, and in the absence of mass flow $\rho_n v_n + \rho_s v_s = 0$, hence

$$p = \frac{\rho_n W^2}{\rho_s \rho T^2 S^2} = \frac{W^2}{c_2^2 \rho C T}, \tag{5.11}$$

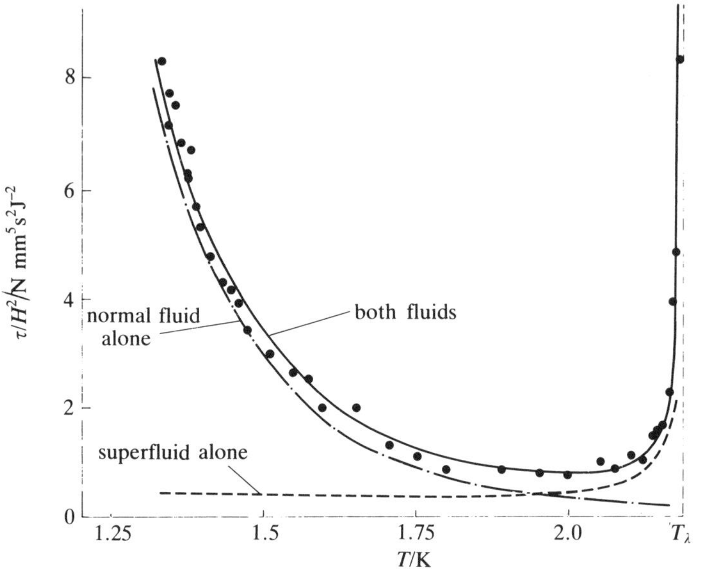

Fig. 5.4. The torque τ on a Rayleigh disc due to an alternating heat flux H, plotted as τ/H^2 versus temperature. The points are experimentally determined; the lines show the theoretical contribution from the normal and superfluid, and from both fluids together (Pellam and Hanson [48]).

where c_2 is the velocity of second sound, ρ and C the density and specific heat of helium, and T the absolute temperature. Equation (5.11) has been verified for a uniform flow of heat, and also by observing the torque due to an oscillating flux of second sound impinging on a disc suspended from a torsion fibre. Because the reaction pressure depends on the square of the heat flux, there is a net torque on the disc even in an oscillating field, just as on a Rayleigh disc in the path of an ordinary sound wave.

The very characteristic temperature dependence of expression (5.11) is shown in Fig. 5.4, which also gives the contribution of each fluid to the total pressure. We see that the torque on a disc rises rapidly at both high and low temperatures. This comes about because either ρ_n or ρ_s is becoming very small. Hence in order to satisfy the condition for no net mass flow of helium

$$\rho_n \boldsymbol{v}_n + \rho_n \boldsymbol{v}_s = 0,$$

either $\boldsymbol{v}_n$ or $\boldsymbol{v}_s$ must become large. Hence, as the torque or pressure depends on the square of the velocities, the net effect of a *decrease* in, say, ρ_n is to *increase* the contribution the normal fluid makes to the pressure.

6
ELEMENTARY EXCITATIONS IN HELIUM II

The basic concept necessary for an understanding of helium II and the two-fluid model is that of the *elementary excitation.* Above the lambda point, liquid ^{4}He behaves rather as a dense classical gas, but at lower temperatures its behaviour is quite different. This is by no means surprising, as the de Broglie wavelengths of the helium atoms are comparable with the interatomic spacings. Therefore, as was first pointed out by Landau [44], it is necessary to describe the atomic motion in terms of elementary excitations.

We have already introduced the concept of elementary excitations when discussing the properties of liquid ^{3}He. We recall that in the Landau theory of a Fermi liquid the atoms were replaced by quasi-particles or elementary excitations. The dispersion relation for these excitations was a form of the single-particle relation $\epsilon = p^2/2m$, modified to take account of the interactions between the atoms. At first sight we might expect the excitations in ^{4}He to be rather similar, but the experiments we now describe show that the excitation spectrum is quite different. In fact, the properties of helium II below 0.6 K imply that the only thermal motion consists of phonons similar to those in crystalline solids. It will therefore be useful to summarize some well-known properties of phonons in such solids.

6.1 Phonons in crystalline solids

The standard treatment of the thermal vibrations of the atoms in crystalline solids is to resolve their motion into normal modes, which are plane waves occupying the whole volume of the solid. On quantizing these normal modes, and taking account of the periodicity of the lattice, we find that the energy of a quantum, or *phonon,* is related to its wave number by a relation $\epsilon = \epsilon(\boldsymbol{k})$ known as the energy spectrum or dispersion curve. This may be determined experimentally by observing the inelastic scattering of slow neutrons, and a typical result is shown in Fig. 6.1.

From the dispersion curve, we can deduce all the thermodynamic

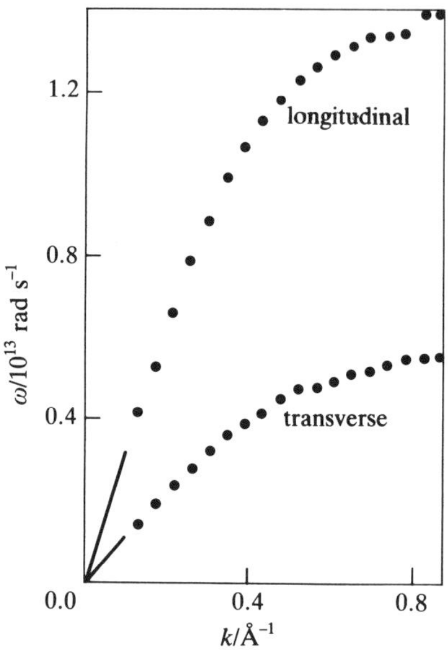

Fig. 6.1. The dispersion curve for transverse and longitudinal waves of wave number k and circular frequency ω in lead at 80 K (Stedman *et al.* [49]).

properties of the lattice using statistical mechanics, provided that the phonons are in a state of thermal equilibrium. In fact, this equilibrium is assured by the small anharmonic terms in the equations of motion of the atoms, which induce transitions between the normal modes. These transitions, besides establishing equilibrium, may also set a limit to the usefulness of our representation of thermal motion in terms of excitations. Thus, if the mean time between transitions is Δt, the energy of a state E is only sharp to within an energy ΔE given by the uncertainty principle

$$\Delta E\, \Delta t \sim \hbar.$$

Clearly, if an excitation is to be a meaningful concept, then ΔE must be considerably less than the energy E. However, this condition is generally well satisfied for phonons in solids.

We are often concerned with non-equilibrium steady-state conditions, as for example in the flow of heat down a steady temperature gradient. We then need to specify how the state of the system is different at different points in space. The phonons so far described are plane waves occupying the whole volume of a solid, and therefore do not readily take account of spatial variations. We deal with such problems by making up localized excitations from wave packets of plane waves, which travel through the solid with the group velocity of the waves. A flow of heat between the hot and cold ends of a rod arises because the density of phonons is greater at the hotter end, and hence there is a larger flow of phonons towards the cold end than away from it.

The above picture of heat flow leads to the well-known expression for the thermal conductivity of a dielectric solid

$$K = \tfrac{1}{3}\mathscr{C} c\lambda,$$

where $\mathscr{C}$ is the specific heat per unit volume, c the velocity of sound, and λ the mean free path of the phonons between scattering processes. The conductivity is thus proportional to the specific heat, the velocity of sound, and the mean free path which takes account of the various scattering processes which impede the free flow of the phonons. In a pure dielectric material at room temperature the only scattering processes are between the phonons themselves, that is each phonon is scattered by the density variations due to other phonons.

Phonons interact with each other mainly by three-phonon processes of two types. *Normal processes* conserve energy and wave-vector, so that

$$\epsilon_1 + \epsilon_2 = \epsilon_3 \quad \text{and} \quad \boldsymbol{k}_1 + \boldsymbol{k}_2 = \boldsymbol{k}_3$$

and the flow of energy remains unchanged. On the other hand, *umklapp processes* do not conserve wave-vector; hence they scatter the phonons and produce a thermal resistance. The numbers of both types of interaction decrease rapidly with falling temperatures, so that eventually the mean free path is limited by the walls of the specimen, and the apparent conductivity is then proportional to the diameter of the specimen. This is the so-called size effect associated with boundary scattering.

6.2 Phonons in helium II

Below 0.6 K, the specific heat of liquid ^{4}He varies at T^3. This behaviour is quite different from that of a perfect gas of Bose particles with an energy spectrum $\epsilon = p^2/2m$, which would have a specific heat proportional to $T^{\frac{3}{2}}$ (see for example reference [50]). Moreover, the magnitude of the specific heat of liquid ^{4}He is almost exactly that given by Debye's theory for the specific heat of a *crystalline solid,* provided that we take only longitudinal modes into account, as the liquid cannot support shear waves. Thus, substituting the measured velocity of sound into the usual Debye formula, we obtain

$$C_V = 0.0207 \times T^3\,\mathrm{J\,g^{-1}\,K^{-1}},$$

which is the same as the measured value, to within the experimental error of 2 per cent.

It is remarkable that the specific heat of a liquid may be calculated by assuming that the atomic motion is no more complex than in a solid. Nevertheless the assumption is supported by the form of the melting curve. As discussed in Section 1.2, the vanishing of $\mathrm{d}p/\mathrm{d}T$ implies that the entropy of the liquid along the melting curve is very close to that of the solid. In addition, the thermal conductivity of helium II below 0.6 K strongly suggests that the thermal motion may be completely described in terms of phonons.

Figure 6.2 shows the thermal conductivity of the liquid, measured in two capillaries of diameter 0.29 and 0.80 mm. Below 0.6 K, the apparent conductivity varies as T^3, and is proportional to the diameter of the capillary. This behaviour is reminiscent of the heat flow in a dielectric crystal which is limited by boundary scattering. In fact the results of Fig. 6.2 may be represented by the usual formula for boundary scattering

$$K \simeq \tfrac{1}{3}\mathscr{C}cd,$$

where $\mathscr{C}$ is the specific heat per unit volume, c the velocity of sound, and d the diameter of the specimen.

To sum up, both the specific heat and thermal conductivity of helium II below 0.6 K suggest that the only modes of heat motion are longitudinal phonons analogous to those in a solid. Above 0.6 K the specific heat

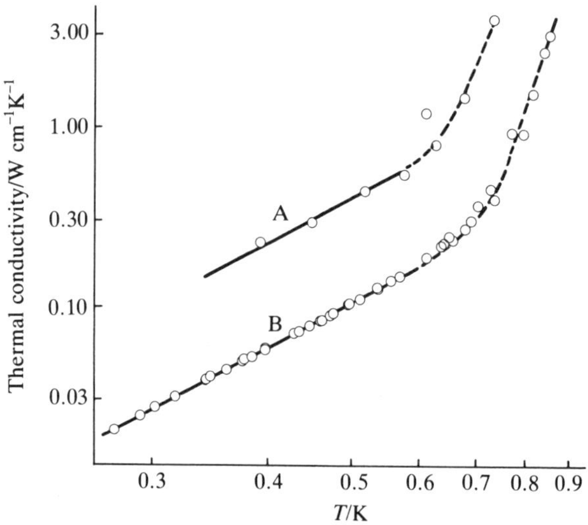

Fig. 6.2. The thermal conductivity of liquid ^{4}He: A, in a tube 0.80 mm diameter; B, in a tube 0.29 mm diameter (Fairbank and Wilks [51]).

varies approximately as T^6, and rapidly increases above the Debye value, which implies that other modes of motion are being excited. The nature of these other modes, and the reason why they are not excited at lower temperatures, form one of the central problems of liquid helium II.

6.3 The scattering of neutrons by helium II

The possibility of detecting the thermal excitations in liquid helium II was first pointed out by Feynman and Cohen [52]. If a beam of monochromatic slow neutrons passes through a specimen of liquid helium, some neutrons will be scattered out of the beam, and the only scattering processes at all probable are ones in which the neutron creates or annihilates a single thermal excitation.

If we perform a neutron scattering experiment on an ordinary liquid using a monochromatic incident beam, then the neutrons scattered in a given direction exhibit a wide range of energies. However, in helium II the heat motion is much reduced, and it is found that neutrons scattered in the same direction have almost the same energy. Consider an incident neutron of wave vector $\boldsymbol{q}_i$ which creates an excitation in the liquid of wave vector $\boldsymbol{k}$ and energy $\epsilon(\boldsymbol{k})$, and then emerges with wave vector $\boldsymbol{q}_s$ at an angle θ to its original direction. Applying conservation of energy and wave vector we obtain the values of k and $\varepsilon(k)$ for one particular excitation, *viz.*,

$$\epsilon(k) = \frac{\hbar^2}{2m}(q_i^2 - q_s^2)$$

and

$$k^2 = q_i^2 + q_s^2 - 2q_i q_s \cos\theta.$$

Hence by observing the values of q_s as a function of θ we obtain sets of values of ϵ and k for the elementary excitations, which are used to trace out the dispersion curve. (The energy of the scattered neutrons varies somewhat because the thermal excitations interact with each other. Hence, they have a finite lifetime τ, and an energy uncertainty given by the uncertainty principle $\Delta\epsilon \sim \hbar/\tau$. This effect is small below 1 K, but becomes more marked towards the lambda point. Eventually, above the lambda point, the scattered neutrons have the wide spread of energy characteristic of an ordinary liquid.)

Figure 6.3 shows the dispersion curve for the excitations in helium II obtained by neutron scattering, the energies being given as degrees Kelvin, and momenta in units of wave number. For small wave numbers the curve is approximately linear, with a slope almost equal to the velocity of sound, the excitations in this region being phonons specified

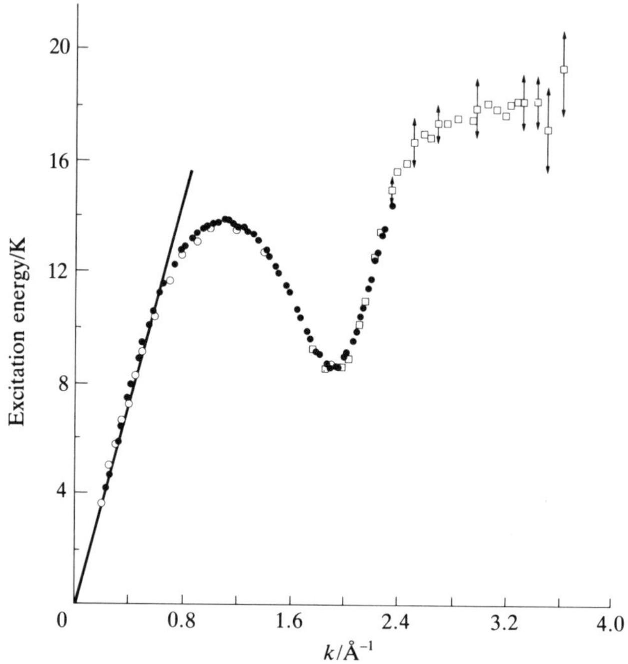

Fig. 6.3. The dispersion curve for the excitations at 1.1 K (after Cowley and Woods [53]).

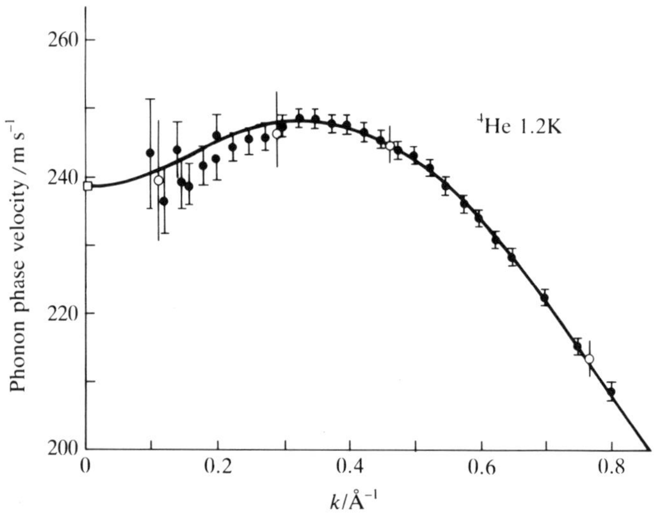

Fig. 6.4. The dispersion curve at low wave number, plotted as $\epsilon/\hbar k$ (the phonon phase velocity) versus wave number k (after Stirling [54]).

by the relation

$$\epsilon = c\hbar k. \tag{6.1}$$

However, even this part of the dispersion curve is not exactly linear, as shown in Fig. 6.4 which gives recent accurate neutron scattering measurements plotted as $\epsilon/\hbar k$ (the phonon phase velocity) vs. k. The observed departure from linearity, though small, is important in that it permits 3-phonon collision processes which would otherwise be forbidden by the conservation of energy and wave vector. We discuss the nature of the excitations of higher momenta in Chapter 11.

Finally we note one important difference between phonons in liquid helium and phonons in a solid. For phonons and other excitations in helium II we will make use of the standard wave-mechanical relation between wave-vector and momentum, namely $\boldsymbol{p} = \hbar \boldsymbol{k}$. However, the normal modes of a crystal lattice are set up by resolving the atomic motions into purely vibrational states plus a translational motion *which is the same for all the atoms*. Hence any momentum of the crystal is associated only with the translation motion, and not with the vibrational modes or phonons. In this case, the quantity $\hbar \boldsymbol{k}$ is not a momentum, although it is sometimes termed a 'quasi-momentum'. In a liquid, the position is quite different. Thus in neutron scattering experiments with a crystal, the lattice is fixed to the spectrometer table, which takes up the recoil of the scattered neutron. A liquid has no rigid lattice, so the recoil momentum is taken up by the individual excitations which have a real momentum $\boldsymbol{p} = \hbar \boldsymbol{k}$.

6.4 Excitations and thermal properties

The dispersion curve permits us to calculate the thermodynamic properties of the liquid. The free energy is given by standard statistical methods as

$$F = \frac{Vk_\mathrm{B}T}{2\pi^2} \int_0^\infty \ln\{1 - \exp(-\epsilon_k/k_\mathrm{B}T)\} k^2 \,\mathrm{d}k. \tag{6.2}$$

At temperatures below 0.5 K, the only appreciable contribution comes from the linear part of the spectrum, and the specific heat

$$C_\mathrm{ph} = \frac{2\pi^2 k_\mathrm{B}^4}{15\rho\hbar^3 c^3} T^3,$$

which is the usual Debye expression for a continuum.

Above 0.5 or 0.6 K excitations of higher momentum are excited. In this case, the thermodynamic functions have to be evaluated by numerical computation. There is also the complication that the form of the

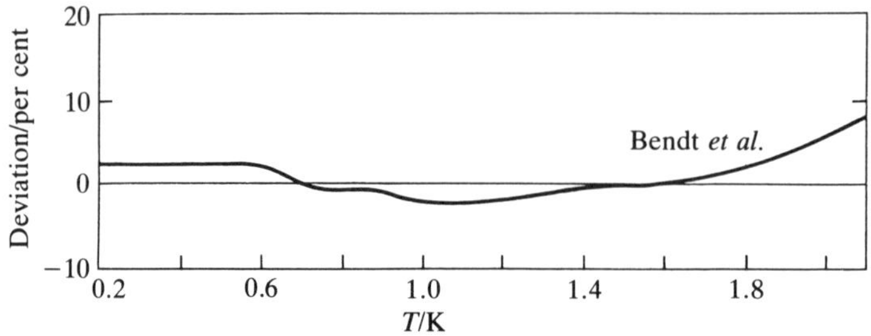

Fig. 6.5. The full line shows the percentage deviation between values of the entropy calculated by Bendt *et al.* [55] and values determined experimentally.

dispersion curve depends slightly on the temperature, but we do not pursue this point further. Figure 6.5 shows the good agreement of the entropy calculated via eqn (6.2) with values determined from the measured specific heat, at least to temperatures of about 1.8 K. (At higher temperatures, the number of excitations becomes so large that the interactions between them must be taken into account.)

6.5 Phonons and rotons

If one takes ϵ_k from the measured dispersion curve (shown in Fig. 6.3) and substitutes into the integrand of eqn (6.2), one finds that only two ranges of k-values are important because of the effect of the Boltzmann factor $\exp(-\epsilon_k/k_B T)$. At the lowest temperatures all the excitations are phonons corresponding to the initial linear part of the dispersion curve. As the temperature is raised, excitations in the region of the minimum are created and cause the entropy to rise rapidly above the Debye value. At all temperatures, the entropy arises principally from the phonons and the excitations near the minimum, although at the higher temperatures there are also contributions from other parts of the dispersion curve. The excitations near the minimum were named *rotons* by Landau, and he represented the dispersion curve in this region by an expression of the form

$$\epsilon = \Delta + \frac{(p-p_0)^2}{2\mu}, \tag{6.3}$$

where Δ and p_0 are the energy and momentum coordinates of the minimum and μ is a constant related to the curvature at the minimum, with values [53]

$$\begin{aligned} \Delta/k_B &\simeq 8.67\ \mathrm{K} \\ p_0/\hbar &\simeq 1.94\ \text{Å}^{-1} \\ \mu &\simeq (0.15)m_4 \end{aligned} \tag{6.4}$$

k_B being Boltzmann's constant, and m_4 the mass of a ^{4}He atom. We discuss the nature of the roton excitations in Chapter 11, but note here that any similarity between the words roton and rotation must be treated very cautiously.

By substituting (6.3) into eqn (6.2) we obtain the contribution of the rotons to the thermodynamic functions. For example the roton entropy

$$S_r = \frac{2(k_B\mu)^{\frac{1}{2}} p_0^2 \Delta}{(2\pi)^{\frac{3}{2}} \rho \hbar^3 T^{\frac{1}{2}}} \left(1 + \frac{3k_B T}{2\Delta}\right) e^{-\Delta/k_B T} \tag{6.5}$$

and the phonon entropy

$$S_{ph} = \frac{2\pi^2 k_B^4}{45 \rho \hbar^3 c^3} T^3. \tag{6.6}$$

At temperatures below 1.3 K where the excitations lie mostly in the phonon and roton parts of the spectrum, measurements of $S(T) = S_{ph}(T) + S_r(T)$ can be analysed to yield values of Δ and the product $\mu^{\frac{1}{2}} p_0^2$. A similar analysis of measurements of the normal density using equation (7.10) leads to values of Δ and the product $\mu^{\frac{1}{2}} p_0^4$. Hence we may determine the values of Δ, p_0 and μ independently of the neutron scattering experiments.

7
THE THEORETICAL BASIS OF THE TWO-FLUID MODEL

We now show how the two-fluid model may be deduced from the properties of the elementary thermal excitations. We begin by giving Landau's discussion of superfluidity and the density of the normal fluid.

7.1 Superfluidity

To maintain the flow of an ordinary liquid through a capillary, it is necessary to apply a pressure, and most of the work done by this pressure appears as heat in the liquid. Without going into any details of the viscous processes, we can say that the resistance to flow arises because of a conversion of some of the kinetic energy of the liquid into disordered thermal motion. Or, put another way, the relative motion of liquid and capillary sets up additional thermal motion in the liquid, and this is the source of the viscous reaction. We now show that excitations can only be formed in helium II, as the result of flow past a wall, if the helium has a certain minimum velocity. If this velocity is not exceeded, there is no energy dissipation arising from the flow, and no viscous loss.

Consider a mass of liquid helium at absolute zero passing through a tube or capillary with streamline flow. How will it respond to the reactions of the wall which tend to slow it down? The slowing down process can only be brought about by the gradual excitation of internal motions in the liquid in the layers close to the wall.

Let us suppose that helium is flowing through a tube which is so constrained that it may move freely in the direction of its length but in no other direction. We regard the flow through the tube from a coordinate system in which the helium is at rest, and the tube moving with velocity $\boldsymbol{v}$ and momentum $\boldsymbol{P} = M\boldsymbol{v}$ where M is its mass. Let us further suppose that the relative motion of tube and liquid creates an excitation in the liquid with energy ϵ and momentum $\boldsymbol{p}$. If the tube and the helium form an isolated system, then the tube will lose an amount of energy

$$\Delta E = \epsilon. \tag{7.1}$$

The tube will also lose momentum $\Delta\boldsymbol{P}$ directed along its length, and from the conservation of momentum it follows that

$$|\Delta \boldsymbol{P}| \leqslant p, \tag{7.2}$$

the equality sign being valid if the excitation moves parallel to the length of the tube. In this case the energy of the tube will be reduced by an amount which can be expressed in terms of the initial and final momenta as $(\boldsymbol{P}_i - \boldsymbol{P}_f) \cdot (\boldsymbol{P}_i + \boldsymbol{P}_f)/2M$. But $\boldsymbol{P}_i$ and $\boldsymbol{P}_f$ are very nearly equal since the excitation's momentum is much smaller than either, so to a good approximation $(\boldsymbol{P}_i + \boldsymbol{P}_f)/2M = \boldsymbol{v}$, and

$$\Delta E = v\,|\Delta \boldsymbol{P}|, \tag{7.3}$$

hence

$$v \geqslant \epsilon/p. \tag{7.4}$$

That is, an excitation can only be created as the result of the flow if the velocity is at least ϵ/p.

Landau's relation (7.4) shows that a certain minimum velocity of flow v_c is needed to create an excitation in the moving helium. In the initial linear phonon region of the spectrum $\epsilon = cp$ so that v_c is equal to the velocity of sound ($\sim 2 \times 10^4$ cm s^{-1}). The minimum value of ϵ/p for any excitation is seen by inspection of the dispersion curve (Fig. 6.3) to occur in the region of the minimum (roton excitations) and to be approximately equal to Δ/p_0 (6×10^3 cm s^{-1}). Hence, provided no other internal motions are possible, superfluidity should persist up to these high velocities.

Experiments with negative ions (see Section 10.6) have observed the predicted critical velocity in situations where the only dissipative mechanism is the creation of pairs of rotons. In most experiments, however, the critical velocities are much lower than the 60 m s^{-1} predicted for the creation of rotons, being generally of the order of a few cm s^{-1} or less. This discrepancy comes about because Landau assumes that the flow of the liquid is streamline, and thus rules out the possibility of other forms of internal motion besides the phonons and rotons. In fact we describe in Chapters 13 and 14 experiments which show that vortex-like motions are set up by relatively low velocities of flow. We defer until that chapter all discussion as to the value of the relevant critical velocity.

Although the observed critical velocities are often less than that predicted by Landau's treatment, his discussion of the creation of thermal excitations is *essential* to explain why the liquid is superfluid. To appreciate this point, let us consider the flow of an ordinary liquid through a tube. In this case, the thermal excitations are excitations of single atoms, and the dispersion curve has the form $\epsilon = p^2/2m$. Thus, for sufficiently small values of p, there will always be possible excitations for which ϵ/p is less than any specified velocity. Therefore excitations can always be created by flow, and the liquid is not a superfluid. The superfluidity of helium II arises from the difficulty of creating the particular forms of thermal excitations which are permitted in the liquid.

7.2 The Galilean transformation

Before proceeding further, we collect here one or two riders concerning the Galilean transformation between two coordinate systems with relative velocity v, as shown in Fig. 7.1 The relation between the two sets of coordinates is

$$\boldsymbol{r}' = \boldsymbol{r} + \boldsymbol{v}t \quad \text{and} \quad t' = t.$$

It follows directly from classical mechanics that the energies (E', E) and momenta $(\boldsymbol{P}', \boldsymbol{P})$ of a body of mass M as viewed from the two systems are related by the equations

$$E' = E + \boldsymbol{P} \cdot \boldsymbol{v} + \tfrac{1}{2}Mv^2,$$
$$\boldsymbol{P}' = \boldsymbol{P} + M\boldsymbol{v}.$$

We will need expressions relating the energies and momenta of an excitation viewed from these two coordinate systems. We therefore consider a mass M of helium at temperature zero, containing no excitations, moving with velocity $\boldsymbol{v}$ relative to the laboratory coordinates. To an observer moving with the liquid, the helium has zero momentum and an energy E_0. To an observer at rest in the laboratory, the helium will have a momentum and energy given by the above transformations, namely $M\boldsymbol{v}$ and $E_0 + \frac{1}{2}Mv^2$ as set out in Table 7.1. Let us now suppose that an excitation is created in the liquid, so that the observer moving with the liquid sees a total momentum $\boldsymbol{p}$ and a total energy $E_0 + \epsilon$. The total momentum and energy as seen by the laboratory observer are again given by the Galilean transformation, and are $\boldsymbol{p} + M\boldsymbol{v}$ and $E_0 + \epsilon + \frac{1}{2}Mv^2 + \boldsymbol{p} \cdot \boldsymbol{v}$ respectively. These results are displayed in Table 7.1. We

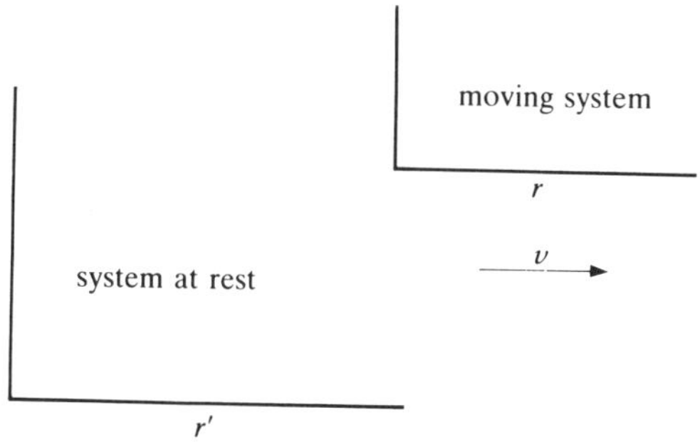

Fig. 7.1. Two coordinate systems with relative velocity v; see text.

Table 7.1

	Momentum		Energy	
	Coordinate system moving with the helium	Laboratory coordinate system	Coordinate system moving with the helium	Laboratory coordinate system
No excitations present	0	Mv	E_0	$E_0 + \frac{1}{2}Mv^2$
One excitation present	$\boldsymbol{p}$	$\boldsymbol{p} + M\boldsymbol{v}$	$E_0 + \epsilon$	$E_0 + \epsilon + \frac{1}{2}Mv^2 + \boldsymbol{p} \cdot \boldsymbol{v}$

see that, as far as the laboratory observer is concerned, the extra momentum and energy associated with the excitation are $\boldsymbol{p}$ and $\epsilon + \boldsymbol{p} \cdot \boldsymbol{v}$. Hence the relation between the energy appropriate to the moving coordinate system and to the laboratory system (indicated by primes) is

$$\epsilon' = \epsilon + \boldsymbol{p} \cdot \boldsymbol{v}. \tag{7.5}$$

We also see that the momentum (or wave number) is invariant to the relative velocity of two systems, that is

$$\boldsymbol{p}' = \boldsymbol{p}. \tag{7.6}$$

We now consider how the distribution function must be modified when the excitations have an imposed drift velocity $\boldsymbol{v}_n$ relative to the observer. To fix our ideas, let us commence with the helium at rest, in equilibrium, and in the same coordinate system as the observer. The appropriate distribution function is the usual expression for an unspecified number of Bose particles

$$n_0(\epsilon) = \{\exp(\epsilon/k_B T) - 1\}^{-1}. \tag{7.7}$$

We then impose a drift velocity $\boldsymbol{v}_n$ on the excitations. For an observer moving with this velocity the distribution is still given by (7.7). That is, he sees a certain number, $n_0(\epsilon)$, of excitations with energy ϵ. However, to a stationary observer these excitations appear to have an energy $\epsilon + \boldsymbol{p} \cdot \boldsymbol{v}_n$, in accord with eqn (7.5). Hence for an observer at rest

$$n_0(\epsilon + \boldsymbol{p} \cdot \boldsymbol{v}_n) = \{\exp \epsilon(p)/k_B T - 1\}^{-1}$$

or, what amounts to the same thing,

$$n_0(\epsilon) = \{\exp(\epsilon - \boldsymbol{p} \cdot \boldsymbol{v}_n)/k_B T - 1\}^{-1}. \tag{7.8}$$

7.3 The normal density

The frictionless flow of helium II at absolute zero comes about because of the difficulty of creating excitations mechanically. We now consider the behaviour of the liquid when it flows through a capillary at higher temperatures, when some thermal excitations are already present. The argument used to derive the critical velocity did not depend on the fact that the helium was in its ground state. Hence no new excitations will be produced by flow at higher temperatures unless the same critical velocities are exceeded. However, the existing excitations will collide with the walls of the tube, and momentum will be exchanged between the tube and the liquid.

Consider a mass of helium II in a long, straight tube of negligible thermal capacity, both the tube and the helium being at rest and at temperature zero. The temperature of the helium is then raised so as to create excitations thermally. These excitations have energies and momenta, ϵ and $\boldsymbol{p}$, which are the quantities measured in neutron scattering experiments. The tube is next moved relative to the helium in a direction along its length. Because of collisions with the tube, the excitations will eventually come into equilibrium with a drift velocity equal to that of the tube, say $\boldsymbol{v}_n$.

Let us now calculate the total momentum associated with the excitations in a volume V of liquid helium. By symmetry, the net momentum of the excitations $\boldsymbol{P}$ must be in a direction parallel or anti-parallel to $\boldsymbol{v}_n$. Also, the density of states in p space is $4\pi V p^2 \,\mathrm{d}p/h^3$, and of these states a fraction $\frac{1}{2}\sin\theta\,\mathrm{d}\theta$ corresponds to values of $\boldsymbol{p}$ making angles with $\boldsymbol{v}_n$ in the range θ to $\theta + \mathrm{d}\theta$. Hence the net momentum of the excitations is

$$\boldsymbol{P} = \hat{\boldsymbol{v}}_n \int_0^\infty \int_0^\pi p\cos\theta \cdot n' \cdot \frac{4\pi V p^2}{h^3}\,\mathrm{d}p \cdot \tfrac{1}{2}\sin\theta\,\mathrm{d}\theta,$$

where n' is the actual (non-equilibrium) value of the distribution function. Expanding n' about its equilibrium value in powers of $\boldsymbol{v}_n$, neglecting second order and higher terms, and integrating over θ, we obtain

$$\boldsymbol{P} = \rho_n V \boldsymbol{v}_n,$$

where

$$\rho_n = -\frac{4\pi}{3h^3}\int \frac{\partial}{\partial\epsilon}\{\exp \epsilon/k_B T - 1\}^{-1} p^4\,\mathrm{d}p. \tag{7.9}$$

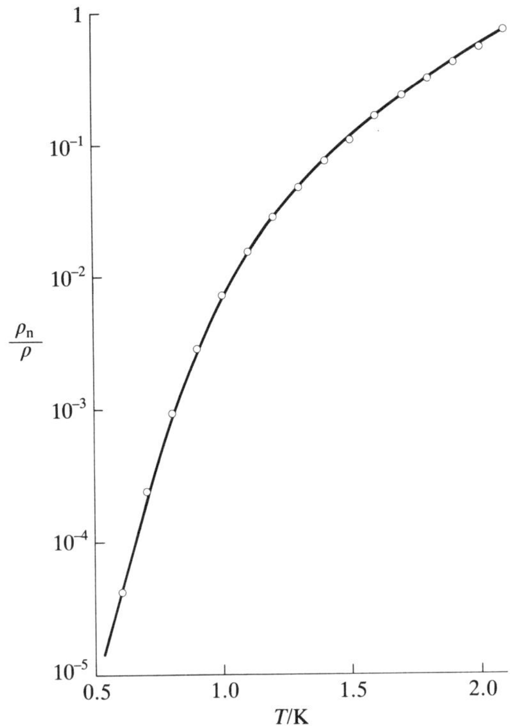

Fig. 7.2. Values of ρ_n/ρ calculated by Brooks and Donnelly [56] from neutron scattering data. The circles represent values given by Wilks [57] which were derived from second sound velocities measured by Peshkov [41]. Data obtained by Andronikashvili (see Fig. 5.2) also fit well and a comparison with other data is given in [56].

Note that ρ_n is an essentially positive quantity with the dimensions of density, and is a function of temperature, but not of $\boldsymbol{v}_n$.

Values of ρ_n derived from eqn (7.9) and the dispersion relation given by the neutron scattering experiments are shown by the full line in Fig. 7.2. Note that ρ_n approaches the total density ρ at the lambda point, and becomes very small at low temperatures. By substituting into (7.9) the dispersion relations for the phonon and roton parts of the excitation spectrum, we find that the phonon and roton contributions to ρ_n are

$$\rho_{nph} = \frac{2\pi^2 k_B^4}{45\hbar^3 c^5} T^4,$$
$$\rho_{nr} = \frac{2\mu^{\frac{1}{2}} p_0^4}{3(2\pi)^{\frac{3}{2}}(k_B T)^{\frac{1}{2}}\hbar^3} e^{-\Delta/k_B T}. \tag{7.10}$$

The two contributions are equal at about 0.5 K, but ρ_{nph} increases much less rapidly with rising temperature than ρ_{nr}, and therefore contributes less than 1 per cent of the total value above about 1.2 K.

If the helium were an ordinary liquid it would eventually come into equilibrium with the moving tube and have a momentum $\rho V \boldsymbol{v}_{\mathrm{n}}$. In fact the only momentum transfer is to the excitations, and the above argument shows that the equilibrium momentum is $\rho_{\mathrm{n}} V \boldsymbol{v}_{\mathrm{n}}$, where $\rho_{\mathrm{n}} \leqslant \rho$. Therefore the reaction of the liquid on the tube is less than for a normal liquid, and as far as the tube is concerned the helium appears to have a density of only ρ_{n}. This is the type of situation encountered in Andronikashvili's experiment (Section 5.5) and we therefore identify ρ_{n} with the density of the normal fluid in the two-fluid model. Figure 7.2 shows that the calculated values are in good agreement with those derived from experiments on second sound.

7.4 The two-fluid model

We now generalize the argument of the previous section. Suppose that liquid helium at temperature zero is contained in a tube, and that both helium and tube are moving relative to the observer in the laboratory with a velocity $\boldsymbol{v}_{\mathrm{s}}$. The temperature is then raised so as to create excitations. The energy and momentum of an excitation as seen by an observer travelling with the liquid are ϵ and $\boldsymbol{p}$, related by the dispersion curve of Fig. 6.3; but to an observer in the laboratory these quantities appear to have values

$$\epsilon' = \epsilon + \boldsymbol{p} \cdot \boldsymbol{v}_{\mathrm{s}} \qquad \boldsymbol{p}' = \boldsymbol{p} \tag{7.11}$$

in accord with the transformations (7.5) and (7.6). The tube is now given a steady velocity relative to the laboratory of $\boldsymbol{v}_{\mathrm{n}}$ so that when equilibrium is reached the *excitations* have a drift velocity of $\boldsymbol{v}_{\mathrm{n}}$. The appropriate distribution function for the excitations as viewed by a laboratory observer is now given by a generalization of eqn (7.8), namely

$$\begin{aligned} n'(\boldsymbol{p}) &= [\exp\{(\epsilon + \boldsymbol{p} \cdot \boldsymbol{v}_{\mathrm{s}} - \boldsymbol{p} \cdot \boldsymbol{v}_{\mathrm{n}})/k_{\mathrm{B}}T\} - 1]^{-1} \\ &= [\exp\{(\epsilon - \boldsymbol{p} \cdot \boldsymbol{v}_{\mathrm{n}} - \boldsymbol{v}_{\mathrm{s}})/k_{\mathrm{B}}T\} - 1]^{-1}. \end{aligned} \tag{7.12}$$

This expression is similar to (7.8), save that $\boldsymbol{v}_{\mathrm{n}}$ is replaced by $\boldsymbol{v}_{\mathrm{n}} - \boldsymbol{v}_{\mathrm{s}}$. Hence we may proceed as in the previous section, and find that the net momentum of the excitations is now

$$\boldsymbol{P} = \rho_{\mathrm{n}}(\boldsymbol{v}_{\mathrm{n}} - \boldsymbol{v}_{\mathrm{s}})V,$$

where ρ_{n} is defined as previously. The total momentum of the liquid as viewed by the laboratory observer before the creation of the excitations was $\rho V \boldsymbol{v}_s$. Therefore the total momentum per unit volume when the excitations are present is

$$\boldsymbol{j} = \rho \boldsymbol{v}_s + \rho_n(\boldsymbol{v}_n - \boldsymbol{v}_s). \tag{7.13}$$

We now define a new parameter ρ_s by the relation

$$\rho = \rho_n + \rho_s, \tag{7.14}$$

and then rewrite equation (7.13) as

$$\boldsymbol{j} = \rho_n \boldsymbol{v}_n + \rho_s \boldsymbol{v}_s. \tag{7.15}$$

The momentum of the system is now formally resolved into the two components of the two-fluid model.

To complete our derivation of the two-fluid model, we consider the total rate of change of the energy and momentum of the excitations in a unit volume fixed in the laboratory system of coordinates. Changes in this energy and momentum arise

(i) by a change in the number of excitations as the result of flow through the boundaries of the volume,

(ii) by a change in the energy and momentum of the excitations originally present, arising from changes in the density of the helium ρ and from changes in the velocity $\boldsymbol{v}_s$.

Hence, by writing down equations for the energy and momentum balance, we obtain the entropy conservation equation

$$\frac{\partial(\rho S)}{\partial t} = -\mathrm{div}(\rho S \boldsymbol{v}_n) \tag{5.5}$$

and

$$\rho_n \frac{\partial}{\partial t}(\boldsymbol{v}_n - \boldsymbol{v}_s) = -\rho S \;\mathrm{grad}\; T,$$

which is equivalent to equation (5.6) of the two-fluid model (after substituting for grad p from (5.4)).

Finally, as a check to see that our method of calculation is physically meaningful we now calculate ρ_n for a perfect gas obeying Boltzmann statistics. Replacing (7.7) by the Boltzmann distribution, and proceeding as before, we find that $\rho_n = \rho$, the total density, as we would expect. Similarly $\rho_n = \rho$ for a perfect gas obeying Fermi statistics, and for a perfect gas obeying Bose statistics at temperatures above the condensation temperature. However, for a perfect Bose gas below the condensation temperature, $\rho_n = \rho(N_c/N)$ where (N_c/N) is the fraction of molecules in the condensed state.

The last result clearly suggests that Bose statistics are very relevant to the existence of the two-fluid model. Yet liquid helium is certainly not a perfect gas. The transition temperature in the liquid is depressed as the pressure is increased, whereas in a perfect gas the condensation temperature rises. The specific heat of a Bose gas varies as $T^{\frac{3}{2}}$ at the lowest temperatures, whereas liquid helium has a specific heat varying as T^3. We return to this point in Chapter 15.

7.5 Second sound

In this section we consider a relatively simple interpretation of the propagation of second sound or temperature waves [58]. The simplest case occurs when the only excitations are phonons. The Boltzmann equation for the phonon distribution function $n(x_i, p_i)$ may then be written, using tensor notation, as

$$\frac{\partial n}{\partial t}+\frac{cp_i}{p}\frac{\partial n}{\partial x_i}=I_n \qquad (p=|p_i|),$$

where we assume that all the phonons travel with the same velocity c, and I_n is the rate of change of n due to collisions. Assuming conservation of energy and momentum in the collisions, it follows that

$$\frac{\partial E}{\partial t}+c^2\frac{\partial p_i}{\partial x_i}=0 \quad \text{and} \quad \frac{\partial p_i}{\partial t}+c\frac{\partial T_{ik}}{\partial x_k}=0,$$

where

$$E=\int ncp\,\mathrm{d}^3p \qquad p_i=\int np_i\,\mathrm{d}^3p \qquad T_{ik}=\int\frac{p_ip_k}{p}n\,\mathrm{d}^3p.$$

Assuming also that the distribution function is disturbed only slightly from spherical symmetry, $T_{ik}\simeq\frac{1}{3}\delta_{ik}E/c$ so that

$$\frac{\partial^2E}{\partial t^2}=\frac{c^2}{3}\frac{\partial^2E}{\partial x_i^2}. \tag{7.16}$$

As E is proportional to the energy density, eqn (7.16) is the equation of a temperature wave propagating with a velocity $c/\sqrt{3}$.

An extension of the above method for the more general case of excitations with a spectrum $\epsilon=f(p)$ leads to a wave velocity given by

$$c_2^2=\frac{\rho TS^2}{\rho_n C}, \tag{7.17}$$

which differs from the usual two-fluid expression by a factor ρ_s/ρ. However, at temperatures below 1 K this factor is unity to within 1 per cent or less, so that (7.17) gives virtually the same velocity as the standard formula. In the low-temperature limit when only phonons are present (7.17) leads to the velocity $c_2 = c/\sqrt{3}$ predicted by (7.16). Thus at the lower temperatures we may picture second sound as a fluctuation in the distribution function, propagating as a harmonic wave in the gas of excitations. At higher temperatures the picture is more complicated because an appreciable mass of the liquid is associated with the excitations. (This is taken into account by the factor ρ_s/ρ.)

For this model of second sound to be valid, two conditions are essential. Firstly there must be sufficient interaction between the excitations, so that a local equilibrium can be established within distances which are small compared with the wavelength of the sound. That is, the mean free path of the excitations between collisions must be appreciably less than the wavelength of the sound. Estimates of the mean free path may be derived from the theory of the normal viscosity η_n, described in the next chapter. It turns out that above 1 K the mean free path is much smaller than the wavelength in any of the experiments so far reported. However, the mean free path increases rapidly with falling temperature because the number of excitations decreases. Hence at lower temperatures it eventually becomes difficult to obtain signals of second sound. Several authors have investigated the propagation of second sound below 1 K using the pulse method; as the temperature is lowered, the pulse travels faster through the liquid in accord with the two-fluid expression of Section 5.3, but its shape changes considerably. This effect is shown in Fig. 7.3, which is based on photographs of pulses displayed on an oscilloscope. At a temperature of 1.17 K the transmitted pulse is well defined, and characteristic of second sound. At 0.75 K the pulse has moved to the left, nearer to the start of the time base, and is appreciably wider; by 0.26 K the trace is so different that we cannot speak of a second-sound pulse at all.

The second condition for the propagation of second sound is that collisions between the excitations must conserve energy and momentum (or, more accurately wave-vector). For example, if any umklapp-type collisions (cf. Section 6.1) were to occur in helium II, second sound would only be possible if the mean free path between such collisions were much longer than the wavelength of the sound. In fact there are no umklapp processes in helium II, as the liquid has no regular lattice structure. Hence this second condition is always satisfied.

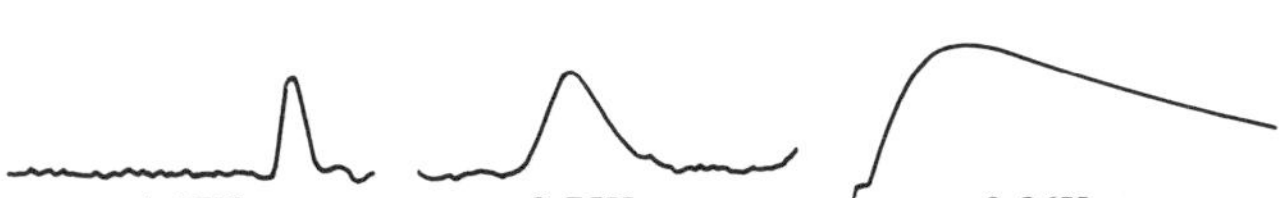

Fig. 7.3. The shape of heat pulses transmitted through helium II at various temperatures as viewed on an oscilloscope trace; for details see text (after Kramers [59]).

Finally we note that this treatment of second sound should apply to any substance in which the only excitations are phonons; hence we might expect to observe temperature waves in dielectric solids. From what we have said above, second sound will only be observed if

$$\lambda_N \ll l \ll \lambda_U, \tag{7.18}$$

where λ_N and λ_U are the mean free paths between normal collisions which conserve wave-vector, and between umklapp collisions which do not, and where l is the wavelength of the sound. Equation (7.18) shows that to observe second sound we must select a crystal in which N processes are much more frequent than U processes. In principle this should be possible as the number of N processes varies as a power law of the temperature, whereas the number of U processes varies exponentially with temperature. Hence, by going to sufficiently low temperatures, λ_N will always be smaller than λ_U. However, in most crystals, the value of λ_N is then so large that the phonons are scattered diffusely by impurity atoms and by the walls of the specimen. Nevertheless, second sound has been observed in crystals of sodium fluoride, *solid* helium, and bismuth (for references see Maris [59a]).

8
KINETIC PROCESSES IN HELIUM II

We described in Chapter 7 how the two-fluid model may be derived by regarding helium II as an assembly of elementary excitations. We now outline how this approach has been extended by Landau and Khalatnikov to account for the coefficient of the normal viscosity, for a coefficient analogous to the thermal conductivity of an ordinary liquid, and for the attenuation of sound and temperature waves.

The above coefficients are estimated by regarding the helium as an assembly of interacting excitations subject to a velocity or temperature gradient, or to an applied periodic disturbance. The calculations are then analogous to those used when computing the thermal conductivity of a dielectric solid. For example, the phonons in a solid diffuse under the influence of a temperature gradient, the actual rate of flow being determined by the number and nature of the collisions between them. The position in helium is similar, but more involved, as there is a larger spectrum of excitations, and several different types of possible collisions. As the calculations are somewhat lengthy, we shall only summarize the main results.

8.1 The normal viscosity

As mentioned in Section 6.1, interactions between excitations will transfer a system from one state to another, giving a finite lifetime to each state. One such interaction involving a total of five excitations might be written schematically as

$$E_1 + E_2 + E_3 \rightleftharpoons E_4 + E_5.$$

We see that although energy is conserved, the number of excitations may or may not be. As in Section 6.1, we regard the excitations as being localized, and talk of the interactions between them as collisions; this is a convenient representation though not a necessary one.

Collisions will occur between excitations of all wave numbers, and therefore the evaluation of the collision integral is very involved. To simplify the problem, Landau and Khalatnikov [60] note that below about 1.5 K the excitations are predominantly those which can be represented by either the phonon or roton spectrum. They therefore confine their attention to phonons and rotons. The most probable types of collisions are in general those involving the smallest number of excitations . In fact the behaviour of the liquid can be accounted for by considering the following processes (not all of which are equally important):

3-phonon	$P_1 \rightleftharpoons P_2 + P_3$
3-roton	$R_1 \rightleftharpoons R_2 + R_3$
4-phonon	$P_1 + P_2 \rightleftharpoons P_3 + P_4$
4-roton	$R_1 + R_2 \rightleftharpoons R_3 + R_4$
5-phonon	$P_1 + P_2 \rightleftharpoons P_3 + P_4 + P_4$
P–R Scattering	$P_1 + R_1 \rightleftharpoons P_2 + R_2$
P–R Conversion	$P_1 + R_1 \rightleftharpoons R_2 + R_3$

By considering the above processes in detail, Landau and Khalatnikov obtain expressions for the various collision cross-sections. The phonon cross-sections depend principally on the velocity of sound and its variation with the density, while the roton cross-sections involve an interaction function which has to be determined from experiment as described below. It is then possible to write down a Boltzmann transport equation for the motion of the excitations, and to calculate the momentum flux when a velocity gradient is imposed on the system. The calculation is complicated as we must integrate over all possible values of the momenta of the excitations, and it is also necessary to decide which of the various collision processes are the most important.

It turns out that the normal viscosity η_n has two components η_r and η_{ph}. The first component η_r is associated with a flux of rotons scattered mainly by rotons. The number of rotons falls exponentially with temperature, but their mean free path increases at about the same rate, so this viscosity turns out to be independent of temperature. The component η_{ph} arises from a flux of phonons which is scattered principally by the rotons. As the liquid is cooled the number of phonons is reduced, nevertheless the viscosity rises because the number of rotons (causing scattering) decreases exponentially. Note that the value of η_r depends on the magnitude of the unknown interaction, but that η_{ph} is completely specified by the theory.

To compare these results with experiment, we restrict ourselves to values of the viscosity below about 1.9 K, because at temperatures near the lambda point the excitations are so numerous that interaction effects become important. Between about 1.5 and 1.9 K the viscosity does not

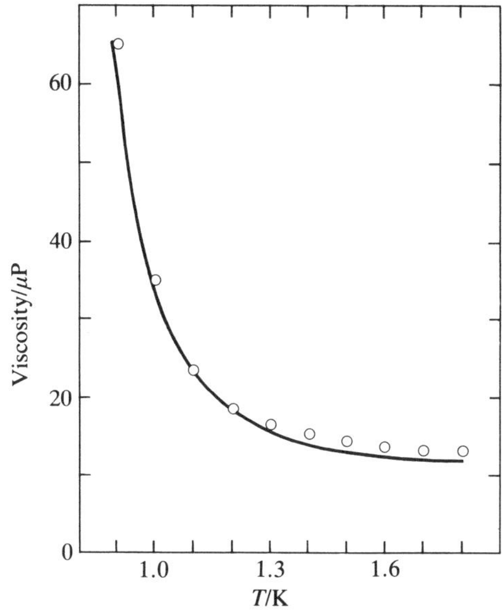

Fig. 8.1. Values of the normal viscosity η_n derived theoretically (Khalatnikov [61]). The points show experimental values.

change much with temperature, and the calculated value of η_{ph} is very small. Landau and Khalatnikov therefore identified the whole of the viscosity in this region with η_r, and wrote $\eta_r = 1.2 \times 10^{-5}$ poise. Then, calculating values of η_{ph}, they derived the total viscosity $(\eta_r + \eta_{ph})$, as shown by the full line in Fig. 8.1. Considering the approximations involved, and the rather scanty information on some of the parameters, the agreement is surprisingly good. Note that when Landau and Khalatnikov wrote their first paper the viscosity had only been measured down to about 1.3 K.

8.2 The thermal kinetic coefficient

The thermal kinetic coefficient K is a measure of the *irreversible* flow of heat due to a temperature gradient. This flow is quite different from that due to the convection of the two fluids specified by eqn (5.9). The convective flow is reversible, and does not lead to any dissipation. It is also much greater than that associated with K, hence the effects of the latter are often hard to observe. The coefficient K has the form and dimensions of a thermal conductivity, and is sometimes described as the thermal conductivity of the normal fluid, but although there is some truth in this description, it is a misleading simplification. For example if only phonons are present, K vanishes completely. Thus K depends in a very complicated way on the interactions between the excitations. The calculations of K proceeds in a similar way to that of the viscosity. The only unknown parameter is the roton interaction function and this may be deduced from the value of the roton viscosity η_r.

We cannot observe the heat transport associated with K in the presence of a steady temperature gradient as it is so much smaller than the transport due to the relative motion of the two fluids (Section 5.7). However, its effect is obvious in the propagation of second sound which consists of oscillations of the two fluids in anti-phase (Section 5.3). This motion is reversible, but there is a dissipation of energy associated with the temperature gradient in the normal fluid, and this leads to absorption of the second sound with a coefficient of absorption

$$\alpha_2 = \frac{2\pi^2 f^2}{\rho c_2^3} \frac{K}{C}, \tag{8.1}$$

where f and c_2 are the frequency and velocity of the second sound, and ρ and C the density and specific heat of the helium. Although other irreversible processes such as the normal viscosity have some small effect, the absorption is associated principally with the coefficient K; and the full line in Fig. 8.2 shows values of this absorption calculated from eqn (8.1)

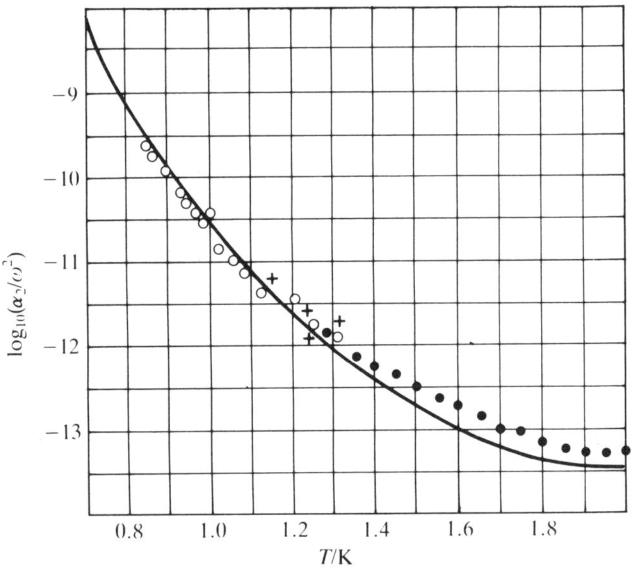

Fig. 8.2. The coefficient of absorption α_2 for second sound of angular frequency ω. ○ Zinov'eva [62]); ● Hanson and Pellam [63]; + Atkins and Hart [64]. The full line shows calculated values. (After Khalatnikov [65]).

and the theoretical values of K. Figure 8.2 also shows three sets of experimental values for the attenuation of waves and pulses of second sound, and we see that there is good agreement between theory and measurement. This may be somewhat fortuitous in view of the various approximations made in the calculation of the coefficient K, and because some of the necessary parameters are not known too accurately. Even so, the agreement gives substantial support to the concepts involved in the coefficient K, and to Khalatnikov's method of evaluation.

8.3 The absorption of sound

The coefficient of the absorption of ordinary sound in liquid ^{4}He varies with temperature in a complex way, as shown in Fig. 8.3. We may distinguish at least three temperature regions in which the absorption arises from quite different mechanisms. Above the lambda point, the absorption is accounted for by the usual processes of viscosity and thermal conductivity as in ordinary liquids. At the lambda point there is a very high absorption associated with the phase transition, and then a large peak in the region of 1 K. (There is also a region below 0.6 K, which we do not consider, where only phonons are present.)

The large peak at about 1 K has been explained by Khalatnikov. In order to discuss a rather complicated situation involving all the excitations, Khalatnikov made the simplifying assumption that in equilibrium there are N_r rotons all having the same energy Δ (the roton minimum) and N_{ph} phonons all with another single energy which is an appropriate

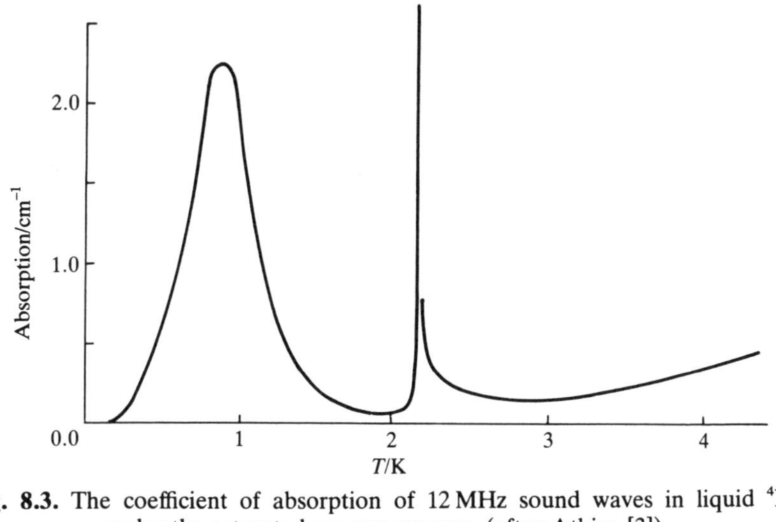

Fig. 8.3. The coefficient of absorption of 12 MHz sound waves in liquid ^{4}He under the saturated vapour pressure (after Atkins [3]).

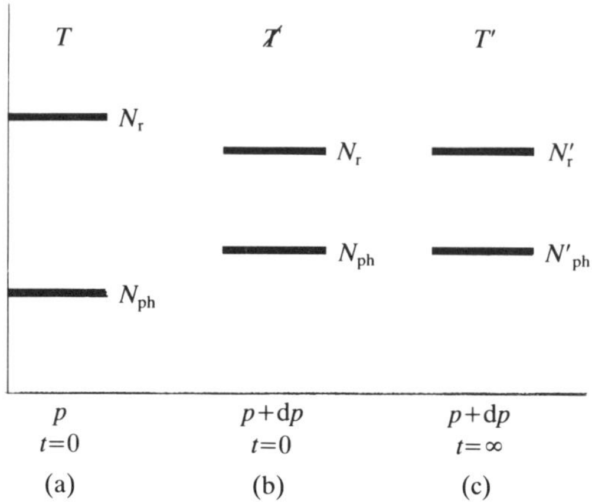

Fig. 8.4. Schematic diagram of the changes in the energy levels and occupation numbers which occur when helium II is subjected to an instantaneous pressure change (after Dransfield *et al.* [66]).

average of the spread of actual phonon energies. The absorption comes about as follows. Consider a mass of helium in thermodynamic equilibrium at temperature T and pressure p, and let the energy levels and their occupation numbers be as represented in Fig. 8.4(a). On compressing the liquid the energy levels are displaced to the positions shown in Fig. 8.4(b); the phonon level is displaced upwards and the roton level downwards. (The height of the roton level is given by the parameter Δ, and $(\partial\Delta/\partial\rho)$ is known to be negative from neutron scattering experiments.) If the compression is carried out rapidly, the occupation numbers N_{ph} and N_r remain unchanged, and no longer correspond to the equilibrium state for the new levels. Therefore a redistribution of energy takes place between the different modes, until the liquid is again in internal equilibrium; it will then be at a somewhat different temperature T' characterized by occupation numbers N'_{ph} and N'_r (Fig. 8.4(c)). This redistribution involves the emission and absorption of phonons and rotons, and comes about principally by the five-phonon and phonon–roton processes of Section 8.1. These processes do not happen instantaneously but are characterized by certain relaxation times. Hence, if the liquid is subjected to an alternating pressure, the numbers of excitations will lag behind their equilibrium values; the motion will be irreversible and absorption will occur. This irreversible response of a system to a change in volume is characterized by another kinetic coefficient, the so-called *second viscosity*.

The absorption at a given temperature will vary with frequency in a manner typical of all relaxation processes (Fig. 8.5(a)). That is, it has a

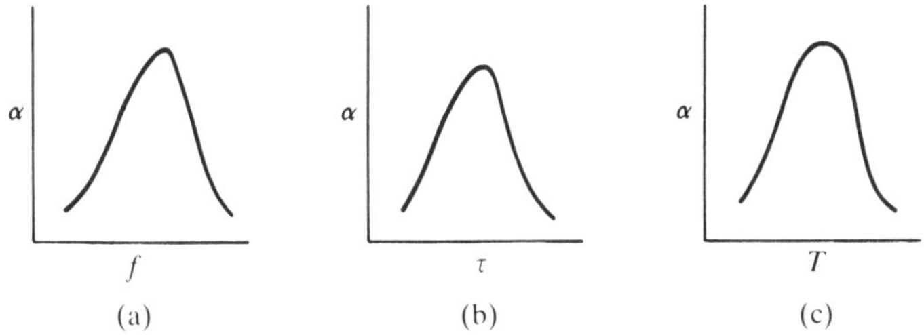

Fig. 8.5. Schematic diagram showing the absorption α due to a relaxation process as a function of frequency f, relaxation time τ, and temperature T.

maximum value when the period of the sound wave is comparable to the relaxation times in the liquid, and is very small at both high and low frequencies. (At low frequencies, the numbers of excitations maintain their equilibrium values; at high frequencies they never change. In both cases the motion is reversible.) One might study this relaxation by working at constant frequency and somehow varying the relaxation times τ within the liquid, so tracing out a curve of the form shown schematically in Fig. 8.5(b). In fact, the relevant relaxation times change very rapidly with temperature. Hence, by working at constant frequency, and varying the temperature, we obtain the form of curve shown in Fig. 8.5(c) similar to the peak in Fig. 8.3. Khalatnikov's treatment gives a good semi-quantitative account of the observed absorption, and of its dependence on the frequency of the sound and the density of the helium. The theory has been reviewed more recently by Khalatnikov [45].

8.4 The thermal boundary resistance

In the course of Kapitza's early experiments on heat flow in helium II he attempted to measure the temperature gradient around a heated wire immersed in the bulk liquid. As the apparent conductivity of the liquid is very high, no temperature gradient could be observed in the liquid itself, but the wire ran appreciably hotter than the liquid. Kapitza also showed by an ingenious experiment that this temperature difference was located within a few hundredths of a millimetre of the surface of the wire. At the time of these experiments, the temperature drop was associated with a thermal resistance arising from the anomalous properties of helium II. However, it now appears that this so-called Kapitza resistance is merely one example of the thermal resistance which occurs at a boundary between any two different media. In the general case when the media are both normal solids, perfectly joined, a simple model based on acoustic mismatch [67] predicts the heat flow to be

$$\dot{Q} = \frac{2\pi^5 k_B^4 \Gamma A}{15\hbar^3 c_{hot}^3}\left[T_{hot}^4 - T_{cold}^4\right] \tag{8.2}$$

where k_B is Boltzmann's constant, A the area of the surface, c_{hot} the velocity of sound in the hotter medium, and the dimensionless parameter Γ is calculated from the densities and sound velocities on either side of the interface. It is generally somewhat less than unity. It follows that the thermal resistance of the boundary has the form

$$R_K \text{ (thermal resistivity)} \equiv A\Delta T/\dot{Q} \simeq \text{constant}/T^3. \tag{8.3}$$

Other mechanisms (e.g. spin–phonon, phonon–electron, and phonon–phonon interactions) can be included and the theory becomes quite involved. In some cases with normal solids reasonable agreement between theory and experiment has been found (see for example Harrison [68]). However, if one of the media is liquid helium, either pure ^{3}He or ^{4}He or a mixture, the heat flow is always larger than the predicted value, sometimes by orders of magnitude. Also the temperature dependence may be quite different from that predicted so that eqn (8.3) is then not of an appropriate form, but it is nevertheless frequently used as a standard for comparison.

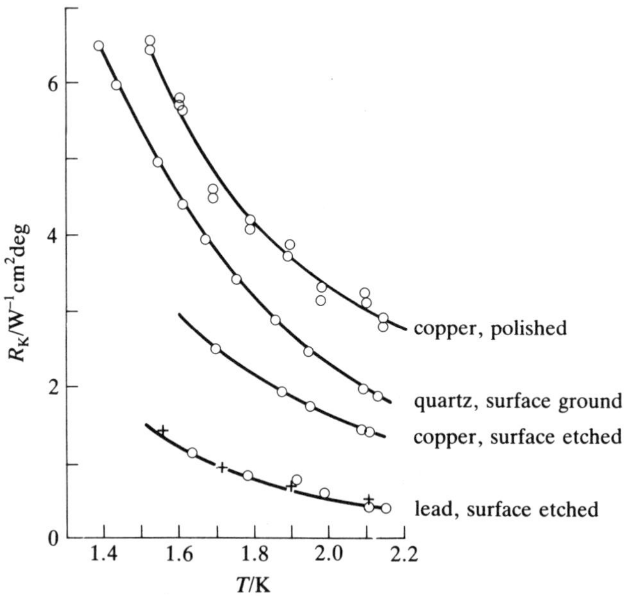

Fig. 8.6. The Kapitza thermal boundary resistivity R_K between liquid helium II and copper, quartz, and lead (Challis *et al.* [69]).

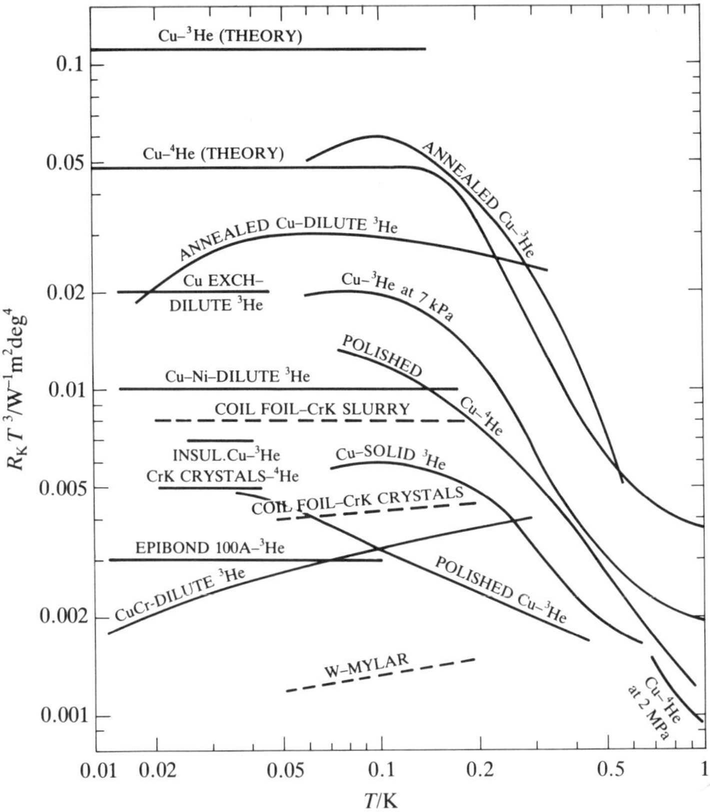

Fig. 8.7. The Kapitza thermal boundary resistivity between liquid helium isotopes and various solids. The dashed lines show some solid–solid resistivities. As suggested by eqn (8.3), the resistivities are plotted as $R_K T^3$ vs. T but it can be seen that the equation's success is very limited. (From Lounasmaa [5]).

It is very fortunate for the experimentalist working at ultra-low temperatures that the heat flow is higher than that given by eqn (8.2), but in many cases the detailed mechanisms are still obscure. The case of superfluid ^{4}HeII is of particular interest because virtually all the temperature drop occurs at the boundary. In other liquids there is a considerable temperature gradient in the liquid, and it is difficult to resolve the boundary resistance from that of the bulk material. Harrison [68] has given a lengthy review, and some typical values of the boundary resistance are shown in Figs. 8.6 and 8.7. The heat flow depends both on the nature of the materials and the state or treatment of the interface. Hence, as it is extremely difficult to specify the state of the surfaces, it is not surprising that the experimental results are not entirely reproducible.

9
SUPERFLUID ^{3}He

When ^{3}He first became available after the Second World War, the principal point of interest was whether liquid ^{3}He would show superfluid properties similar to those of liquid ^{4}He. The first experiments by Osborne *et al.* [70] showed that liquid ^{3}He did not become superfluid down to at least 1 K. Then in 1957 Bardeen *et al.* [71] published their theory of superconductivity, now known as the BCS theory, in which the phenomenon of electron pairing allows conduction electrons (fermions) to form pairs and condense in a way reminiscent of Bose–Einstein condensation. This result led to discussions on the possibility of observing similar pairing in ^{3}He, see for example Anderson and Brinkman [72]. The first estimate made for the transition temperature was about 100 mK but this value was progressively amended to lower values as experiments failed to find any transition. In fact the transition occurs at so low a temperature that its discovery had to await the development of two new methods of reaching low temperatures: the dilution refrigerator described in Section 10.7 which is based on the properties of dilute solutions of ^{3}He in liquid ^{4}HeII, and the Pomeranchuk refrigerator based on the properties of pure ^{3}He. In order to understand what follows we must first describe how this Pomeranchuk refrigerator works.

9.1 The Pomeranchuk refrigerator

The melting curve of ^{3}He exhibits a minimum melting pressure of 29.3 atm at 0.32 K. This implies that below 0.32 K the entropy of the melting solid is greater than that of the freezing liquid at the same temperature. Figure 9.1 illustrates the entropies and the melting curve and shows that substantial cooling can be achieved by a reversible and adiabatic reduction of volume, for example from point I to point F. Starting from the lowest temperatures available with a dilution refrigerator it is only necessary to solidify a fraction of the liquid in order to reach temperatures close to 1 mK. Nevertheless, care has to be exercised in the compression since if even small amounts of the mechanical work are converted into frictional heating the cooling effect can be reduced or even reversed. Precautions are also necessary to avoid blocking the capillary

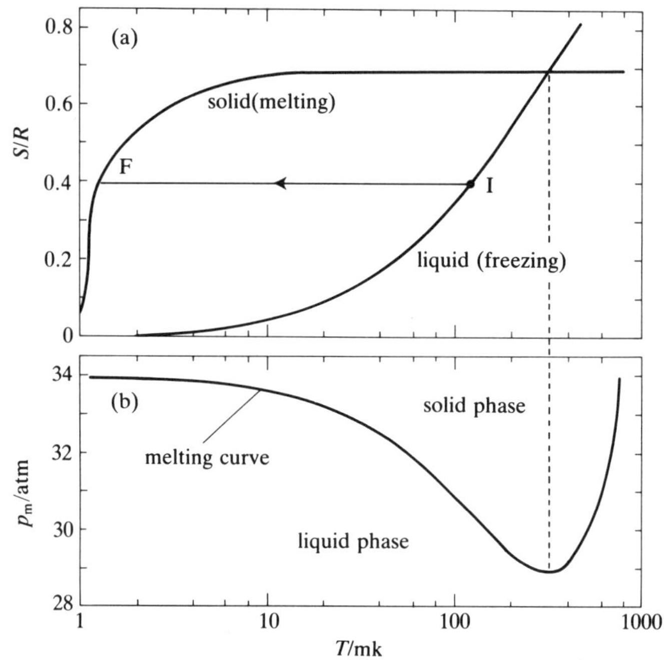

Fig. 9.1. (a) The entropy of freezing liquid and of melting solid ^{3}He and (b) the melting pressure of ^{3}He, as functions of temperature. Cooling from I to F can be achieved by a reversible adiabatic volume reduction controlled by a pressure increase. (From Betts [5]).

when applying the pressure. For example, if pressure is steadily applied to a cell at 0.1 K through the filling capillary, then as soon as the applied pressure reaches the lowest pressure on the melting curve, the capillary will block at the point where the temperature is 0.32 K. As the pressure is raised further the block will spread and isolate the cell from further changes. However this difficulty may be overcome by the use of a double bellows arrangement in which ^{4}He is used to compress the ^{3}He, or by other methods.

9.2 The superfluid phases

In late 1971 Osheroff *et al.* [73] at Cornell were making experiments to seek an ordering transition in *solid* ^{3}He which had been predicted in the region of 2 mK. They were using a Pomeranchuk cell, because of its low operating temperature and insensitivity to magnetic fields, and were making their experiments on the refrigerant ^{3}He itself. Figure 9.2 shows some of their observations as they varied the volume of the cell.

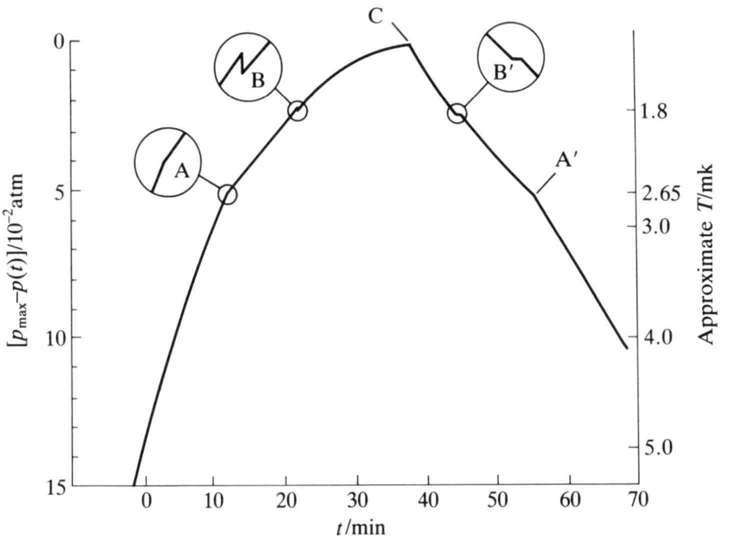

Fig. 9.2. Pressure vs. time trace when the volume in the Pomeranchuk cell was changed at a constant rate; NOTE the unusual labelling of the axes. The features on the trace, including the A and B transitions, are discussed in the text. (After Osheroff *et al.* [73]).

The ^{3}He was compressed over a period of about 35 min in such a way that the volume *decreased* at a *constant rate* until point C was reached at the summit of the curve. The pressure was then steadily reduced so that the volume then *increased* at a *constant rate.* The left-hand vertical axis shows the pressure increasing with time up to point C where $p(t) = p(\max)$, and then decreasing. Because the Pomeranchuk cell contains both liquid and solid phases in equilibrium, the pressure also determines the temperature, which is shown on the right-hand axis. At about 2.65 mK, point A, the slope of the curve changed abruptly by roughly a factor two. At about 1.8 mK, point B, a sudden drop of about 0.3 mbar was observed but with no change in the slope. During the reversed motion of the piston beyond C, the features B′ and A′ were observed. A and A′ were invariably seen at the same temperature whereas B always occurred at a lower temperature than B′, suggesting that some super-cooling occurred during the compression. Moreover, if the rate of volume change was increased, the temperature of B decreased further.

This work was published under the title *Evidence for a New Phase of Solid ^{3}He* [73] but various other workers suggested that the effects could be related to new phases of *liquid* ^{3}He. Within a few months, NMR measurements by Osheroff *et al.* [74] showed magnetic changes in the liquid which correlated with the A and B transitions. The new phases of

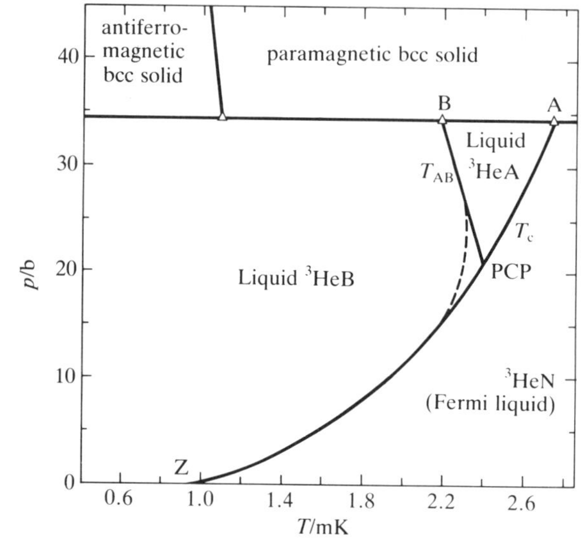

Fig. 9.3. The ^{3}He phase diagram in zero applied field; see text. The effect upon T_{AB} of a 0.0378 T magnetic field is shown by the dashed line. (From Halperin *et al.* [80]).

liquid ^{3}He were thus established. We now give a broad picture of these phases; there are also a number of reviews in the literature, both elementary [75, 76] and more advanced [77–79].

Figure 9.3 illustrates the phase diagram of ^{3}He in zero magnetic field below about 3 mK. The figure includes four labelled points. A and B mark the transitions from ^{3}HeN to ^{3}HeA and from ^{3}HeA to ^{3}HeB along the melting curve; PCP marks the polycritical point at which ^{3}HeN, ^{3}HeA, and ^{3}HeB coexist; and Z marks the zero-pressure transition from ^{3}HeN to ^{3}HeB. The table below gives the pressure and temperature of these four points as quoted by Alvesalo *et al.* [4]; these values are not undisputed but they will suffice for present purposes. Note that curve

Point	p bar	T mK
A	34.4	2.79
B	34.4	2.16
PCP	21.5	2.56
Z	00.0	1.04

A–PCP–Z which gives the temperature–pressure relation $T_c(p)$ for the second-order transition from the normal phase, is smooth at PCP; whereas curve B–PCP which is the temperature–pressure relation $T_{AB}(p)$ for the first-order A to B transition, meets curve A–PCP–Z at a finite

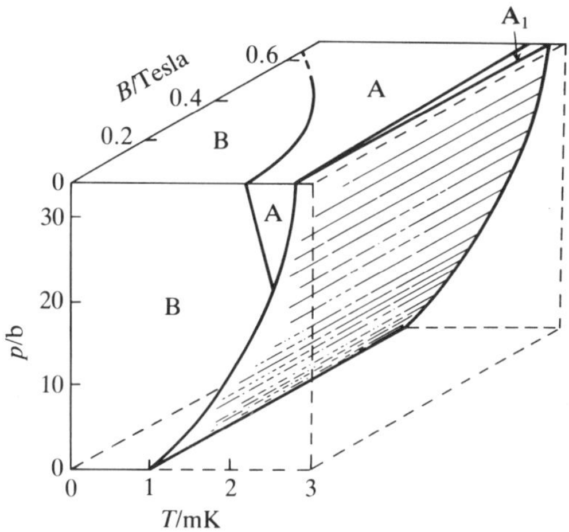

Fig. 9.4. The phase surface in p–B–T space for ^{3}He below 3 mK. The superfluid phases are A, A_1 and B, the normal Fermi liquid region lies to the right of the shaded surface.

angle at PCP. A most unusual feature of the phase diagram is that the application of a small magnetic field has a profound effect as shown by the dashed line in Fig. 9.3. The polycritical point disappears and the B phase can only be reached from the normal phase via a corridor of the A phase at any pressure. This corridor becomes extremely narrow as the temperature falls, for example it is only about 20 μK wide in a field of 0.0378 T at 10 bar [81], and cannot be shown in the figure. As the magnetic field is raised further, the T_c curve alters only slightly whereas the T_{AB} curve is shifted to lower temperatures by an amount proportional to the square of the magnetic field, and the B-phase is entirely suppressed in fields of order 1 T. There is also the further important effect that a magnetic field produces another phase located between ^{3}HeN and ^{3}HeA, and known as ^{3}HeA$_1$. The temperature range over which this A_1 phase is stable at the melting pressure varies approximately linearly with applied magnetic field and extends from zero in zero field to a little over 0.5 mK in about 10 T. These various points are illustrated in Fig. 9.4.

9.3 The specific heat

Liquid ^{3}He, like liquid ^{4}He and superconductors, exhibits a specific heat anomaly centred on the superfluid transition temperature. Figure 1.3 shows the anomaly at the saturated vapour pressure (approximately zero) compared with the lambda point anomaly in ^{4}He, also at the saturated vapour pressure. Figure 9.5 shows some results for ^{3}He under a pressure

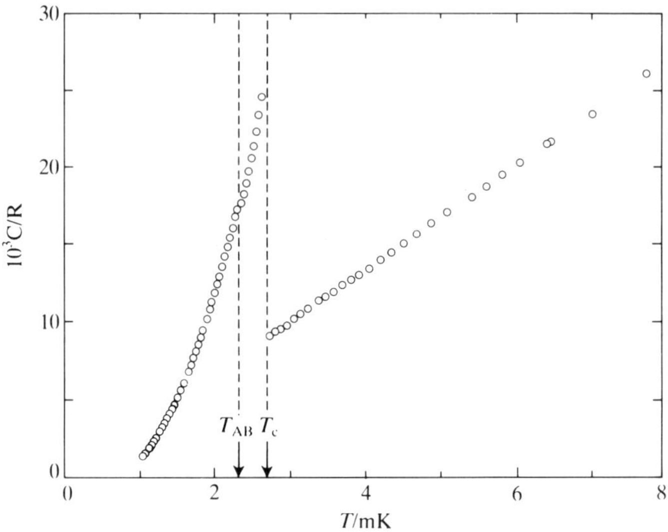

Fig. 9.5. The specific heat of liquid ^{3}He vs. temperature measured at $p = 28.7$ b in zero external magnetic field. The superfluid transition at $T_c = 2.71$ mK and the AB transition at $T_{AB} = 2.34$ mK are indicated by arrows. From Alvesalo *et al.* [4].

of 28.7 bar, the most prominent anomaly is at T_c with a smaller feature noticeable at T_{AB}. It is obvious that these data do not show the almost logarithmic spike of ^{4}He but rather a finite jump similar to that observed in superconductors, and characteristic of a second order transition. As the pressure is raised from zero, the temperature T_c of the transition rises. The magnitude of the specific heat jump ΔC is conveniently expressed in dimensionless form by dividing by the specific heat $C_>$ of the normal phase just above the transition. It is found that $\Delta C/C_>$ rises from a value of 1.4 at $p = 0$ to about 2 at $p = 30$ bar [4]. The other much less prominent feature at T_{AB} is observed only at pressures between p(PCP) and the freezing pressure; this transition is first order with $\Delta C = 0$ and with a latent heat which has been measured [80].

9.4 Superfluid flow

Pekola *et al.* [82] have recently shown that it is possible for superfluid currents of ^{3}HeB to flow apparently undiminished in a hollow torus for at least 48 hours. This implies a viscosity at least 12 orders of magnitude lower than in the normal fluid just above T_c. Their apparatus is shown very diagrammatically in Fig. 9.6, and may be thought of as resembling a steering wheel (the torus) with two different perpendicular pairs of spokes, the wheel being supported by a steering column along its z-axis.

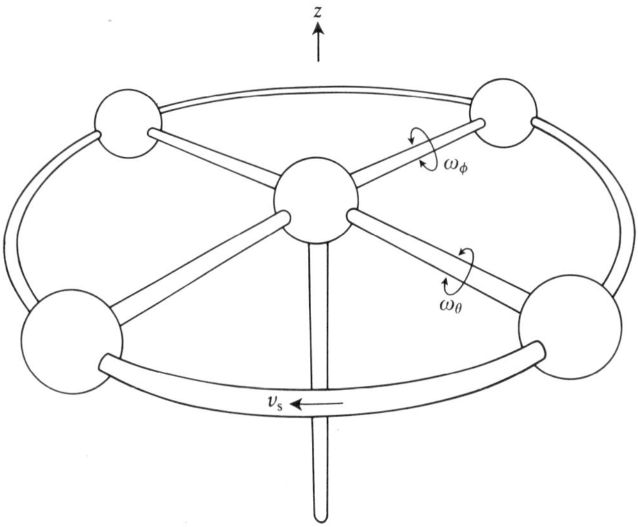

Fig. 9.6. Highly simplified representation of the torus used by Pekola *et al.* [82] in demonstrating superfluidity in ^{3}HeB, see text.

The torus, with $R = 22\,\text{mm}$ and $r = 3\,\text{mm}$, was packed with 20-μm powder in order to raise the critical velocity. The design of the spokes was intended to make the resonant frequency of vibration about the θ axis (ω_θ) much greater than the resonant frequency about the ϕ-axis ($\omega_\phi/2\pi = 69.8\,\text{Hz}$). During an experiment a superconducting solenoid was used to oscillate the torus about the θ axis at the frequency ω_ϕ, and the resulting amplitude of vibration about the ϕ axis was measured using a capacitance bridge. Liquid ^{3}He was induced to flow round the torus at a suitably low temperature in a rotating nuclear demagnetization refrigerator described by Hakonen *et al.* [82*a*]. The flow of liquid produces a gyroscopic effect and its angular momentum comes into the equations of motion and affects the amplitude of the vibrations.

The above demonstration of superfluidity in ^{3}HeB is most convincing but the same authors were unable to demonstrate persistent currents in ^{3}HeA, probably because of what are known as textural effects (see Section 9.9). However, Gammel *et al.* [83] appear to have detected persistent currents in both ^{3}HeA and ^{3}HeB using a rather different technique. Also other authors have made experiments in which liquid ^{3}He was moved through a superleak in a controlled manner with a very thin flexible plastic diaphragm [84]. The rate of transfer and the pressure difference could be found and hence the flow characteristics measured. Flows with undetectable pressure gradients were observed in both ^{3}HeA and ^{3}HeB but these experiments were much less sensitive and do not set

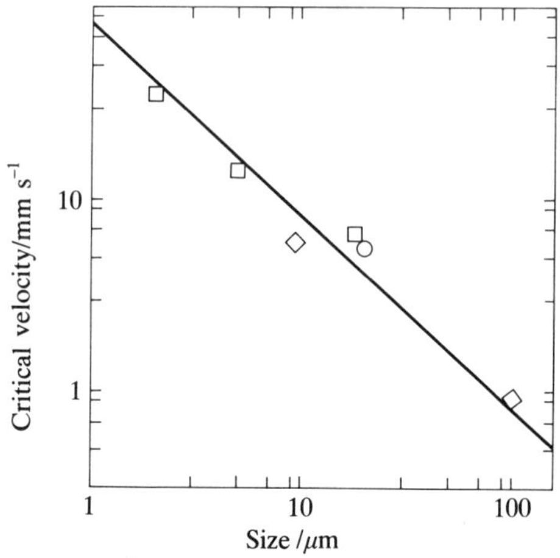

Fig. 9.7. Critical velocity of ^{3}HeB as a function of pore size. The line represents the relationship $v_{sc}d = 6.5 \times 10^{-2}\,\mathrm{mm^2\,s^{-1}}$, given by Gammel *et al.* [85]. The data points are from a variety of sources [82, 85, 86] and have been scaled to 29 b and $T/T_c = 0.5$.

such low limits on the values of the viscosity. Even so, critical velocities v_{sc} were observed and were typically of the order of a few millimetres per second in both phases although not exactly the same in each. The nature of the dissipative mechanisms, of which there seem to be several, have yet to be unambiguously identified. Possible mechanisms include (a) pair-breaking for which v_{sc} is predicted to vary with temperature as $(1 - T/T_c)^{\frac{3}{2}}$, and (b) the creation of vortices or the maintenance of vortex tangles. If vortices are involved, one would expect a dependence of v_{sc} on channel geometry as in ^{4}HeII. Measurements of the dependence on temperature [84] and geometry (Fig. 9.7) have been made but there is still much to be done.

9.5 The two-fluid model

This model, first used to describe ^{4}HeII, has been discussed in earlier chapters. According to this treatment a superfluid is thought of as a mixture of normal (n) and superfluid (s) components with densities ρ_n and ρ_s (with $\rho_n + \rho_n = \rho$) and viscosities $\eta_n \neq 0$ and $\eta_s = 0$. Some of the earliest experiments on ^{3}HeA and ^{3}HeB were designed to investigate whether their behaviour could be described by this model. Kojima *et al.* [87] observed the transmission of pressure waves through both ^{3}HeA and ^{3}HeB in narrow channels where the normal fluid is effectively clamped to the walls by its viscosity. These conditions are similar to those for observing fourth sound in ^{4}HeII, and according to the two-fluid model the velocity of the pressure wave, or fourth sound, is proportional to $c(\rho_s/\rho)^{\frac{1}{2}}$ where c is the velocity of ordinary sound. It was found that the measured values of the velocity implied values of ρ_s which increased steadily as the temperature fell below the transition temperature, as would be expected on the two-fluid model.

Kojima *et al.* only made measurements down to $T/T_c = 0.8$, but the density of the normal component of ^{3}HeB has been measured to lower temperatures in an experiment based on the Andronikashvili method, at pressures between 0 and 29 b [88]. The results show that the normal density tends to zero at the lowest temperatures, again consistently with the two-fluid model (Fig. 9.8). It is seen that ρ_n/ρ is principally a function of T/T_c with relatively small effects from pressure other than the direct effect on T_c. In contrast to ^{3}HeB, ^{3}HeA shows marked anisotropy and ρ_n is not a simple quantity; this is discussed in Sections 9.9 and 9.10. However both phases apparently show a form of behaviour associated with the normal density in the two-fluid model. Leggett [89] has given a theoretical expression for ρ_n in ^{3}HeB:

$$\frac{\rho_n}{\rho} = \frac{[1 + F_1/3]Y}{1 + F_1 Y/3}$$

where F_1 is the normal Fermi liquid parameter (see Chapter 3) related to the effective mass, and Y is the Yoshida function involving T/T_c. Carless *et al.* [90] have given useful series expansions of $F_1(p)$ and $Y(T/T_c)$ which

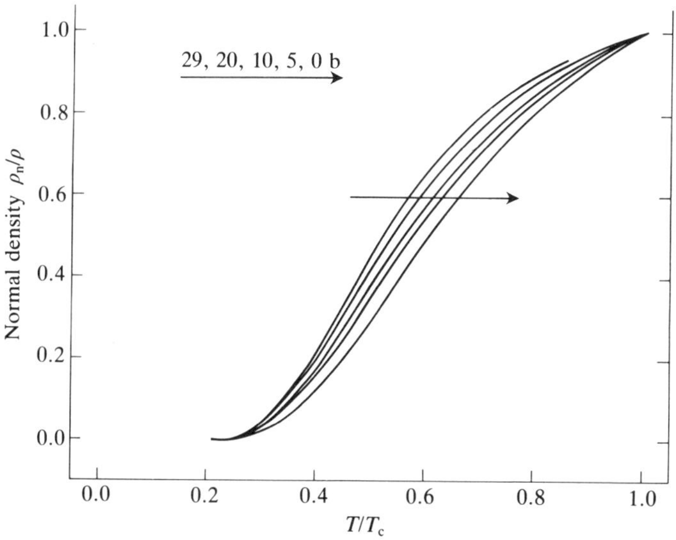

Fig. 9.8. Dependence of ρ_n/ρ on reduced temperature for ^{3}HeB at various pressures. (From Parpia *et al.* [88]).

allow this formula to be conveniently used when numerical values are required.

Other experiments [91] suggest that hydrodynamic heat flow or internal convection takes place in superfluid ^{3}He as in ^{4}HeII (see Section 5.7). However, perhaps the best evidence for the two-fluid model of ^{3}HeA and ^{3}HeB is the recent demonstration [92] that steady heat transfer through either ^{3}HeA or ^{3}HeB is accompanied by hydrodynamic motions (internal convection) which have been detected using a Rayleigh disc as in ^{4}HeII, see Section 5.8. In normal fluids, including ^{3}HeN and ^{4}HeI, heat flow is entirely diffusive and there is no such torque. Hence the torque observed in the superfluid phases appears to be due to the counterflow of the normal and superfluid components.

Figure 9.9 shows values of the viscosity of the normal component of ^{3}HeB measured in various experiments (see Carless *et al.* [90] for details) at pressures close to 20 bar. The theoretical treatment of Ono *et al.* [93] suggests that η_n should show a minimum and then begin to rise slowly as T falls. However, results from several authors (reviewed by Carless *et al.*) which agree well with the theoretical relation above $T/T_c = 0.6$, fall below it at lower temperatures. The discrepancies have been accounted for by Einzel *et al.* [94] in terms of the details of collision processes

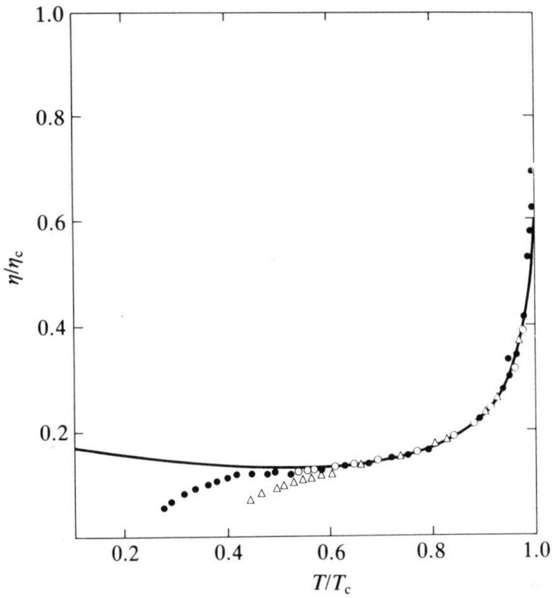

Fig. 9.9. η/η_c in ^{3}HeB as a function of T/T_c measured in experiments at pressures close to 20 b where T_c is about 2.5 mK. (Adapted from Carless *et al.* [90].) The continuous line is given by the theory of Ono *et al.* [93].

occurring at the moving surfaces; these cause the apparent viscosities to differ from the true bulk values in a way which depends on the design of the viscometer.

Measurements of the viscosity and density of the normal fluid in ^{3}HeA are more complicated in that the behaviour of this liquid depends on the direction and magnitude of applied magnetic fields, on the velocity of flow, and on the shape of the container. These result in textural effects, not seen in ^{4}HeII, ^{3}HeN, or ^{3}HeB, which we discuss in Section 9.9. Figure 9.10 shows an example of measurements of ρ_s/ρ for ^{3}HeA in an

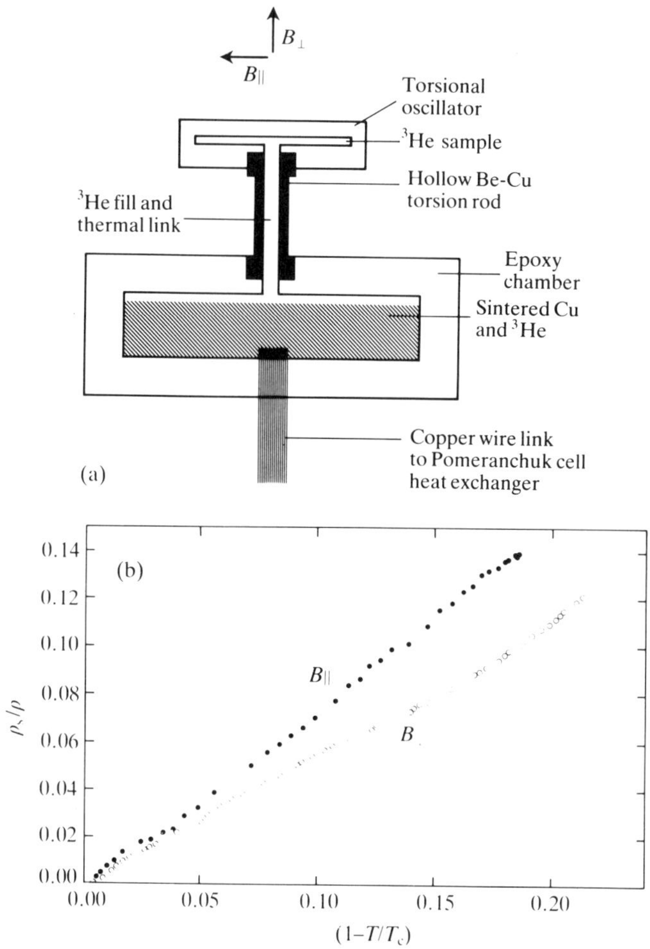

Fig. 9.10. Andronikashvili-type experiment of Berthold *et al.* [95] on ^{3}HeA. (a) The oscillator. (b) The results for ρ_s/ρ in perpendicular and parallel applied magnetic fields of 29 mT.

adaptation of Andronikashvili's experiment; the values depend not only on T/T_c but also on the direction of an applied magnetic field of only 29 mT.

9.6 Nuclear magnetic resonance experiments

Some of the most interesting and important experiments on superfluid ^{3}He deal with its magnetic properties. The ^{3}He nucleus, unlike the ^{4}He nucleus, has a non-zero nuclear spin with quantum number $I = \frac{1}{2}$ giving rise to two possible orientations with respect to a constant applied field B. Its magnetic moment is 2.127 nuclear magnetons ($1.075 \times 10^{-26}\,\mathrm{J\,T^{-1}}$ or $0.778\,\mathrm{mK\,T^{-1}}$). The Larmor angular frequency ω (the rate at which the nuclear spin vector precesses about the direction of an applied field B_0) has the value γB_0 where $\gamma = 2.038 \times 10^8\,\mathrm{rad\,s^{-1}\,T^{-1}}$. Hence when a ^{3}He atom is exposed to radiation at the Larmor frequency, we expect that transitions will be induced between the two levels corresponding to the orientations $I = \pm\frac{1}{2}$. These transitions are observed with $\omega = \gamma B_0$ both in single atoms and in ^{3}HeN, but unusual effects are found in ^{3}HeA and ^{3}HeB.

The experiment of Osheroff, *et al.* [74] on ^{3}HeA, on the melting curve in a Pomeranchuk cell, began with the sample in a steady applied field $\boldsymbol{B}_0$, with a net nuclear spin angular momentum per unit volume represented by the vector $\boldsymbol{S}$. Initially, in equilibrium, $\boldsymbol{S}$ was directed along $\boldsymbol{B}_0$, but then a small extra field was applied perpendicular to $\boldsymbol{B}_0$ in order to jolt $\boldsymbol{S}$ away from $\boldsymbol{B}_0$ by a certain angle; this procedure is known as *tipping* $\boldsymbol{S}$. The subsequent precession of $\boldsymbol{S}$ about $\boldsymbol{B}_0$ was then observed and measured. The results are given in Fig. 9.11 and show that the resonant frequency attributable to the liquid was *not* the Larmor frequency but was shifted from it by an amount which depended on the temperature. Meanwhile the absorption peak attributable to the solid in the cell remained at a fixed frequency, providing a convenient reference point. The area under the liquid peak was constant, from which it follows that the susceptibility was constant. The measured angular frequency $\omega_\perp$ was related to the unshifted value γB_0 in a simple way:

$$\omega_\perp^2 = (\gamma B_0)^2 + \Omega_A^2 \tag{9.1}$$

and experimental values of $\Omega_A(T/T_c)$ are shown in Fig. 9.12.

Another unusual effect shown by ^{3}HeA is called 'parallel ringing' to distinguish it from the more familiar perpendicular ringing which is the precession of $\boldsymbol{S}$ following a tipping pulse as described above. To produce parallel ringing an incremental field $\boldsymbol{\Delta B}$ is suddenly applied *parallel* to $\boldsymbol{B}_0$ and this causes an oscillation of the component of $\boldsymbol{S}$ *parallel* to $\boldsymbol{\Delta B}$ and $\boldsymbol{B}_0$. For small $\Delta B/B_0$ this occurs with a frequency *equal* to Ω_A, that is,

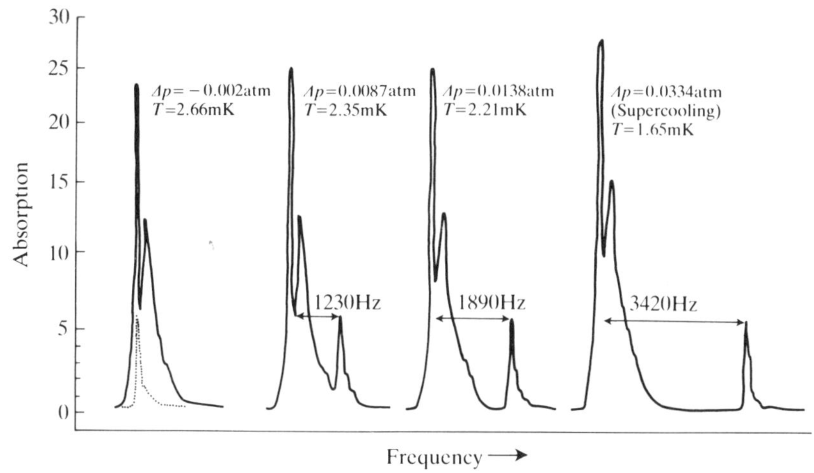

Fig. 9.11. A series of profiles showing that, as the temperature falls, the frequency of the absorption peak due to the A phase of the liquid (on the right) shifts to higher frequencies, away from the double peak due to the solid. The dashed curve shows the unshifted liquid peak at T_c. (From Osheroff *et al.* [74] with the estimated temperatures taken from Lounasmaa [75]).

$\omega_{\|}^2 = \Omega_A^2$ where Ω_A has the same value as in eqn (9.1) describing perpendicular ringing. Some experimental results are plotted in Fig. 9.12. Analogous effects are also observed in ^{3}HeB with a frequency term Ω_B related (with some simplifications) to Ω_A by

$$\Omega_B = \Omega_A(5\chi_B/2\chi_A)$$

where χ_A and χ_B are the magnetic susceptibilities of ^{3}HeA and ^{3}HeB.

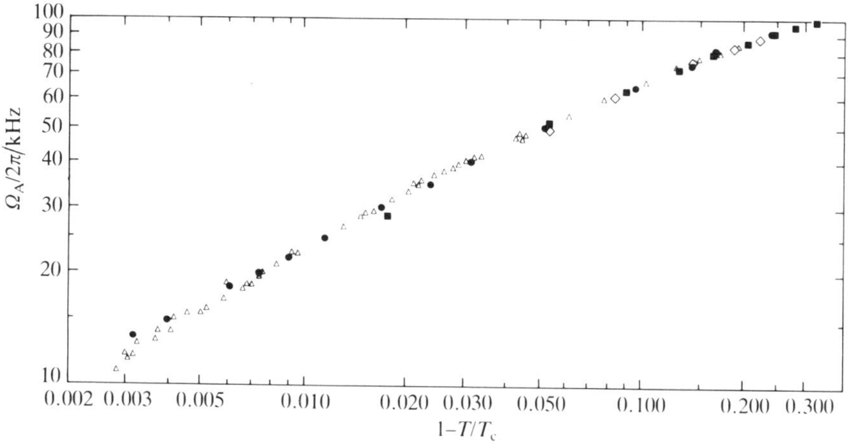

Fig. 9.12. The A phase characteristic frequency $\Omega_A/2\pi$ measured in different ways by different experimental groups. (From Lee and Richardson [78]).

9.7 Pairing and quantum states in superfluid ^{3}He

The results of the various experiments described above can only be understood in terms of correlation effects between individual nuclear magnetic moments as proposed and developed by Leggett [89]. This theory is quite complicated and there is only space here to present some of the basic ideas which are already well established.

The new phases, ^{3}HeA$_1$, ^{3}HeA, and ^{3}HeB are analogous to the superconducting phases in certain metals, the common feature being the pairing of fermions. The transition in superconductors occurs when the formation of electron pairs is energetically favourable. The mechanism is crudely that electron 1 leaves behind it a trail of pulled-in positive lattice ions to which electron 2 is then attracted. The maximum energy benefit would be obtained by completely overlapping trails and this favours an s state for the pair in which the orbital angular momentum quantum number L is zero. For quantum mechanical reasons, the symmetric spatial wave function corresponding to $L=0$ must be combined with an antisymmetric spin state with an even value of the spin quantum number S. Since the two electrons each have half-integral spin, the only possibility is $S=0$ with the electron spins antiparallel.

The pairing of atoms in liquid ^{3}He cannot proceed in quite the same way because the atoms are electrically neutral, and in fact the forces responsible are those between the nuclear magnetic dipoles. A ^{3}He atom with given nuclear spin orientation, say $\uparrow$, induces a cloud of neighbouring spins to favour a particular orientation, say $\downarrow$ (the argument works just as well if the induced orientation is $\uparrow$). The cloud then attracts a second ^{3}He atom with the same orientation as the first, namely $\uparrow$. This attraction between atoms with parallel nuclear spins, and repulsion between atoms with antiparallel spins, results in a bound state with a discrete energy gap between the bound and unbound states (Leggett [89]). Hence, at temperatures sufficiently low compared with the energy gap, pairs form with $S=1$, since each atom has a spin of $\frac{1}{2}$. To obtain the complete wave function of a pair, the symmetric $S=1$ spin state must be combined with an antisymmetric spatial wave function. That is, L must take an odd value, and in the standard model this it taken to be $L=1$ in order to fit with a great deal of experimental evidence. That is, each pair has a non-zero orbital angular momentum.

According to the usual rules of quantum mechanics $S=1$ may have components $S_z=\pm 1, 0$. The three possible spin states $|S, S_z\rangle$ may be represented symbolically as

$$|1, +1\rangle = |\uparrow\uparrow\rangle$$
$$|1, 0\rangle = [|\uparrow\downarrow\rangle + |\downarrow\uparrow\rangle]/\sqrt{2}$$
$$|1, -1\rangle = |\downarrow\downarrow\rangle.$$

These three basis states may be linearly combined in any convenient way to produce a general state $|\Phi\rangle$ and one particular way, which offers some mathematical advantages, is widely used in discussing superfluid ^{3}He. This is to introduce three coefficients d_x, d_y and d_z, and to express the general state $|\Phi\rangle$ as

$$\sqrt{2}\,|\Phi\rangle = d_x[|\downarrow\downarrow\rangle + |\uparrow\uparrow\rangle] + id_y[|\downarrow\downarrow\rangle - |\uparrow\uparrow\rangle] + d_z[|\downarrow\uparrow\rangle + |\uparrow\downarrow\rangle] \quad (9.2)$$

or equivalently as

$$\sqrt{2}\,|\Phi\rangle = [(d_x - id_y)\,|\uparrow\uparrow\rangle + (d_x + id_y)\,|\downarrow\downarrow\rangle + d_z[|\uparrow\downarrow\rangle + |\downarrow\uparrow\rangle]]. \quad (9.2)$$

At first sight either of the equations (9.2) defines a vector $\boldsymbol{d}$, but the position is more complicated because the components of $\boldsymbol{d}$ are not necessarily real, and in fact $\boldsymbol{d}$ is better described as an *order parameter*. It has three important properties. First, the three quantities d_x, d_y, d_z transform as a vector under a rotation. Second, the direction of $\boldsymbol{d}$ is such that its projection along the direction of the spin is zero, that is, $\boldsymbol{d}\cdot\boldsymbol{S}\,|\Phi\rangle = 0$, where $\boldsymbol{S}$ is the net nuclear spin angular momentum per unit volume. (It might appear that this statement only defines the *plane* in which $\boldsymbol{d}$ lies, but the ambiguity is resolved when the complex nature of $\boldsymbol{d}$ is taken into account.) Third, it may be shown [89] that states where $\boldsymbol{d}$ is real have an expectation value of $\boldsymbol{S}$ equal to zero.

The simplest phase appears to be ^{3}HeA$_1$ which exists only in a non-zero magnetic field as shown in Fig. 9.4. It is believed to contain only pairs in the state $|\uparrow\uparrow\rangle$ so the coefficients $(d_x + id_y)$ and d_z must be zero and eqn (9.2) can be expressed simply in terms of d_x only:

$$\sqrt{2}\,|\Phi\rangle = 2d_x\,|\uparrow\uparrow\rangle \qquad (^3\text{HeA}_1).$$

The pairs consist exclusively of atoms whose nuclear spin projections and associated magnetic moment are parallel to the applied magnetic field ($S_z = 1$).

The A phase is believed to conform to the Anderson–Brinkman–Morel model [96], referred to as ABM, in which the wavefunction is taken as a linear combination of the $S_z = \pm 1$ states $|\uparrow\uparrow\rangle$ and $|\downarrow\downarrow\rangle$. As the state with $S_z = 0$ is excluded, the component d_z of the vector $\boldsymbol{d}$ is zero and we can write

$$\sqrt{2}\,|\Phi\rangle = (d_x - id_y)\,|\uparrow\uparrow\rangle + (d_x + id_y)\,|\downarrow\downarrow\rangle \qquad (^3\text{HeA}).$$

The B phase is believed to conform to the Balian–Werthamer model [97], referred to as BW, which takes the general form given in eqn (9.2) above. The ABM and BW models existed before the new phases of liquid ^{3}He had been observed, and the A and B transitions were named arbitrarily. The fact that Anderson's name begins with A and Balian's with B is merely a useful coincidence.

9.8 Liquid ^{3}HeA

The wave-vectors of the two atoms in a pair are $\pm\boldsymbol{k}$ and the net translational momentum of the pair is zero, as in superconductors. The pair forms because its energy is less than that of the two unpaired atoms by an amount known as the *gap* Δ. In superconductors the magnitude of the gap does not depend on the direction of $\boldsymbol{k}$ and is said to be isotropic. The gap in ^{3}HeB is isotropic but in the ABM state of ^{3}HeA, there is at every point in the liquid a preferred axis such that the formation of pairs is energetically most favourable when the orientation of their orbital angular momenta is parallel to the axis. Hence the energy gap Δ is anisotropic, and in fact its magnitude varies as the sine of the angle between the orbital angular momentum $\boldsymbol{l}$ of the pair and the preferred axis.

The direction of the preferred axis is determined by local conditions in the liquid, including for example the proximity of the walls of the container, and may vary slowly from point to point. However all pairs within a small volume will have almost parallel orbital angular momenta, so we can use the vector $\boldsymbol{l}$ to denote both the orbital angular momentum of a *pair,* and to represent a macroscopic vector field in the liquid. In the same way $\boldsymbol{d}$ can be regarded as a property of a single pair and as another macroscopic vector field. A complete discussion of how $\boldsymbol{l}$ and $\boldsymbol{d}$ orient themselves in the presence of outside influences can be very complicated, but if only the magnetic dipole interactions are involved then $\boldsymbol{l}$ and $\boldsymbol{d}$ prefer to be *parallel.*

A purely classical model suggests the above result as follows. Imagine two similar magnetic point-dipoles $\boldsymbol{m}$ attached to the rim of a wheel, one at each end of a diameter, and free to orient themselves in any direction *provided that,* as in the actual ^{3}He pairs, they *both* have to take the same orientation. If the wheel is rotated about its centre fast enough so that the dipoles cannot oscillate, it is merely an exercise in classical electromagnetism to show that the interaction energy, averaged over a whole number of cycles is

$$\frac{3\mu_0 m^2}{8\pi D^3}\cos^2\alpha$$

which has its least value when $\alpha = \pi/2$, where α is the angle between $\boldsymbol{m}$ and the rotation axis of the wheel, and D is the diameter of the wheel (and therefore the separation of the dipoles). So classically the dipoles prefer to lie in a plane perpendicular to the rotation axis, performing a motion rather like a cyclist's feet around the chain wheel of a bicycle. The corresponding quantum mechanical statement is that $\boldsymbol{d}$ and $\boldsymbol{l}$ prefer to be *parallel*; $\boldsymbol{l}$ lies along the wheel axis in our model and $\boldsymbol{d}$ is *perpendicular* to $\boldsymbol{S}$ and to the dipole moment. The energy penalty for non-parallelism is proportional to $\sin^2(\boldsymbol{d}, \boldsymbol{l})$, where $(\boldsymbol{d}, \boldsymbol{l})$ denotes the angle between $\boldsymbol{d}$ and $\boldsymbol{l}$, and hence is proportional to $[1-(\hat{\boldsymbol{d}}\cdot\hat{\boldsymbol{l}})^2]$.

We now see that the resonant frequency in an NMR experiment is no longer equal to γB_0 as in the normal phase because of the mutual interaction of the $\boldsymbol{l}$ and $\boldsymbol{d}$ vectors. As above, the energy penalty for misalignment is proportional to $\frac{1}{2}[1-(\hat{\boldsymbol{d}}\cdot\hat{\boldsymbol{l}})^2]$, and this results in a restoring torque $\lambda(\hat{\boldsymbol{d}}\cdot\hat{\boldsymbol{l}})(\hat{\boldsymbol{d}}\times\hat{\boldsymbol{l}})$ where λ is a function of T/T_c. This torque causes the shift in the resonant frequency described in Section 9.6, an effect which provided the first experimental indication of the existence of the A phase.

The magnitude of the frequency shift may be derived from Leggett's equations of motion for $\boldsymbol{S}$ and $\boldsymbol{d}$:

$$\dot{\boldsymbol{S}} = \gamma\boldsymbol{S}\times\boldsymbol{B} + \lambda(\hat{\boldsymbol{d}}\cdot\hat{\boldsymbol{l}})(\hat{\boldsymbol{d}}\times\hat{\boldsymbol{l}}), \tag{9.3}$$

and

$$\dot{\boldsymbol{d}} = \gamma\boldsymbol{d}\times\boldsymbol{B}_{\text{eff}} = \gamma\boldsymbol{d}\times(\boldsymbol{B}-\mu_0\gamma\boldsymbol{S}/\chi) \tag{9.4}$$

where χ is the susceptibility and the units are SI. The first term of the right hand side of (9.3) is the usual torque experienced by a magnetic moment in an applied field $\boldsymbol{B}$ and the second is the torque tending to align $\boldsymbol{l}$ and $\boldsymbol{d}$. The right hand side of (9.4) expresses the fact that $\boldsymbol{d}$ precesses not about $\boldsymbol{B}$ but about an effective field $\boldsymbol{B}_{\text{eff}}$ in which $\boldsymbol{B}$ is combined vectorially with the field generated by the precession of $\boldsymbol{S}$. Solving these equations leads to expressions for $\boldsymbol{S}(t)$ and $\boldsymbol{d}(t)$ for the experiment of Section 9.6 where $\boldsymbol{S}$ is initially parallel to a constant applied field $\boldsymbol{B}_0$, and is then tipped away from $\boldsymbol{B}_0$ by a small extra field applied perpendicular to $\boldsymbol{B}_0$. The frequency of the subsequent perpendicular ringing, a precession, is predicted by the Leggett equations in the special case of small tipping angles as

$$\omega_\perp^2 = (\gamma B_0)^2 + \gamma^2\mu_0\lambda/\chi \tag{9.5}$$

which is the experimentally observed form, eqn (9.1), with $\Omega_{\text{A}} = \gamma(\mu_0\lambda/\chi)^{\frac{1}{2}}$. The trajectories of the tips of the $\boldsymbol{S}$ and $\boldsymbol{d}$ vectors during this motion are shown in Fig. 9.13(a), while Fig. 9.13(b) and (c) show the trajectories for two larger initial angles of tipping.

The Leggett equations also predict the parallel ringing effect mentioned in Section 9.6, in which, starting from an equilibrium situation with $\boldsymbol{S}$ parallel to $\boldsymbol{B}_0$, a small extra field is applied *parallel* to $\boldsymbol{B}_0$. This does not produce a tipping of $\boldsymbol{S}$ as in the perpendicular case but causes $\boldsymbol{d}$

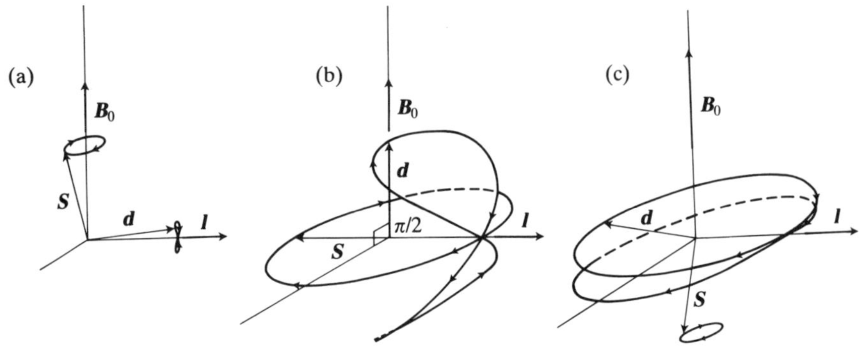

Fig. 9.13. Figures (a), (b) and (c) illustrate the motion of $\boldsymbol{S}$ and $\boldsymbol{d}$ following the tipping of $\boldsymbol{S}$ away from the $\boldsymbol{B}_0$ axis by 10°, 90° and 170° respectively. In each case, the $\boldsymbol{d}$ vector goes through a figure of 8 motion, returning to the axis of $\boldsymbol{l}$ twice in each cycle of the precession of $\boldsymbol{S}$. The rate of precession depends upon the angle between $\boldsymbol{l}$ and $\boldsymbol{d}$ so that the frequency depends on the angle of tipping, as revealed by a full solution of equations (9.3) and (9.4). From Lee and Richardson [78].

to decouple from $\boldsymbol{l}$ and then to oscillate about $\boldsymbol{l}$ in the plane normal to $\boldsymbol{B}_0$ as shown in Fig. 9.14. This oscillation is known as parallel ringing to distinguish it from perpendicular ringing which is a precession. The Leggett equations predict that in parallel ringing the direction of $\boldsymbol{S}$ is unaltered, and that its magnitude varies as

$$\Delta S_z = (\chi/\mu_0\gamma)\,\Delta B(1-\cos\Omega_{\mathrm{A}}t) \tag{9.6}$$

where the frequency Ω_{A} has the same value as in eqn (9.1), $\gamma(\mu_0\lambda/\chi)^{\frac{1}{2}}$.

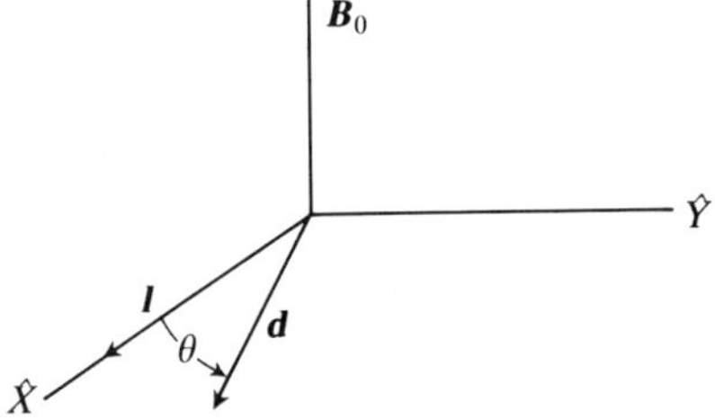

Fig. 9.14. Diagram illustrating the relative orientations of $\boldsymbol{l}$ and $\boldsymbol{d}$ for a parallel incremental field change experiment. From Wheatley [77].

9.9 Textures in ^{3}HeA

In ^{3}HeA in the absence of external influences all the orbital angular momentum vectors of the pairs point in the same direction, giving rise to a macroscopic vector $\boldsymbol{l}$. The nuclear spins are aligned parallel to each other and perpendicular to a second macroscopic vector $\boldsymbol{d}$. However the vectors $\boldsymbol{l}$ and $\boldsymbol{d}$ are also responsive to the various external influences shown in Table 9.1, and there are energy penalties or bonuses associated with these effects. For example, $\boldsymbol{l}$ prefers very strongly to be perpendicular to physical boundaries, i.e. parallel to the normal vector $\boldsymbol{N}$, and $\boldsymbol{d}$ is affected by a magnetic field as in the experiments of the previous Section.

A consequence of these and other influences is that if one visualizes the macroscopic vector fields $\boldsymbol{l}$ and $\boldsymbol{d}$ in ^{3}HeA in the form of 'lines of force', these lines may be curved, and there are further energy penalties incurred in relation to the radii of curvature. In a vessel of given geometry and subject to a given set of electromagnetic and flow fields, the vectors $\boldsymbol{l}$ and $\boldsymbol{d}$ take up patterns or textures which minimize the free energy.

The elegant experiment illustrated in Fig. 9.10 has shown that a measurement of ρ_s/ρ carried out in a modified Andronikashvili oscillator depends sensitively on the orientation of a small applied magnetic field

Table 9.1. ^{3}HeA Free Energy Penalties*

Property	Preferred alignment	Penalty divided by $(1 - T/T_c)$ near T_c, erg cm^{-3}	Magnitude of field to match the magnetic interaction
Magnetic dipole	$\boldsymbol{d} \parallel \boldsymbol{l}$	$-6 \times 10^{-4} (\hat{\boldsymbol{d}} \cdot \hat{\boldsymbol{l}})^2$	
Electric dipole	$\boldsymbol{l} \perp \boldsymbol{E}$	$+2 \times 10^{-12} (\hat{\boldsymbol{l}} \cdot \hat{\boldsymbol{E}})^2$	16 kV cm^{-1}
Susceptibility anisotropy	$\boldsymbol{d} \perp \boldsymbol{B}$	$+5 \times 10^{-7} (\hat{\boldsymbol{d}} \cdot \hat{\boldsymbol{B}})^2$	3.5 mT
Flow	$\boldsymbol{l} \parallel \boldsymbol{v}_s$	$-1 \times 10^{-2} (\hat{\boldsymbol{l}} \cdot \hat{\boldsymbol{v}}_s)^2$	0.25 cm s^{-1}
Wall alignment	$\boldsymbol{l} \parallel \boldsymbol{N}$	$-3 \times 10^{+2} (\hat{\boldsymbol{l}} \cdot \hat{\boldsymbol{N}})$	(very small angle)

* Based on a table given by Lee and Richardson [78].

[95]. Liquid ^{3}HeA was held inside a flat disc-shaped cavity which could be oscillated rotationally about its axis. The viscosity η_n has the effect of locking the normal component (within a finite penetration depth) to the moving surface so its density ρ_n contributes to the moment of inertia, thus altering the mechanical resonance conditions and allowing a measurement of ρ_s. Two textures were investigated (Fig. 9.10). In the first case a small field $\boldsymbol{B}_{\|}$ (29 mT) was applied perpendicular to the axis of the disc. The $\boldsymbol{d}$-vector then prefers to be perpendicular to $\boldsymbol{B}_{\|}$ and parallel to $\boldsymbol{l}$, while $\boldsymbol{l}$ prefers to be perpendicular to the disc faces and parallel to $\boldsymbol{d}$. Both vectors can thus be satisfied by a simple texture in which $\boldsymbol{l}$ and $\boldsymbol{d}$ are parallel to the disc axis. In the second case, $\boldsymbol{B}_{\perp}$ was applied parallel to the axis of the disc. The texture is different and more complicated. At the faces of the discs $\boldsymbol{l}$ must still be parallel to the axis, but $\boldsymbol{d}$ now prefers to be perpendicular to $\boldsymbol{B}_{\perp}$, that is, perpendicular to the axis. It is thus not possible for either $\boldsymbol{d}$ or $\boldsymbol{l}$ to be straight, and the free energy will contain contributions from the magnetic dipole interaction, from the interaction between $\boldsymbol{d}$ and $\boldsymbol{B}_{\perp}$, and from terms due to the curvature of $\boldsymbol{d}$ and $\boldsymbol{l}$. It turns out that $\rho_{n\perp}$ and $\rho_{n\|}$ are different, as shown in Fig. 9.10, and ρ_n cannot be treated as a function only of temperature and pressure.

9.10 Superfluid ^{3}HeA$_1$ and ^{3}HeB

We have so far concentrated on ^{3}HeA for two main reasons. First, many of the most interesting experiments were done first on ^{3}HeA simply because it occurred at relatively higher temperatures than ^{3}HeB and was therefore more easily reached experimentally. Second, the fact that $\boldsymbol{l}$ and $\boldsymbol{d}$ in ^{3}HeA represent both the microscopic properties of a pair and the macroscopic properties of bulk liquid results in some simplification. However, the easiest to understand of the three phases is ^{3}HeA$_1$ because all its nuclear magnetic moments are oriented in the same direction parallel to an applied magnetic field as discussed in Section 9.7.

As mentioned above ^{3}HeA$_1$ only exists in a non-zero field and only in an extremely restricted temperature range of width about 0.5 mK even in an applied field as large as 10 T [98, 99]. As a result of these practical constraints, few of its properties have been measured but Figure 9.15 shows some specific heat results which clearly indicate the existence of the phase. NMR studies have also been carried out, and some of these have established beyond doubt that the nuclear magnetic moments are *parallel* (rather than antiparallel) to the applied field (Ruel and Kojima [100]). Finally, we mention that when ^{3}HeA$_1$ is exposed to a magnetic field gradient, all the pairs experience a force in the same direction, and this enabled Ruel and Kojima to observe an AC superfluid motion by applying such a gradient sinusoidally.

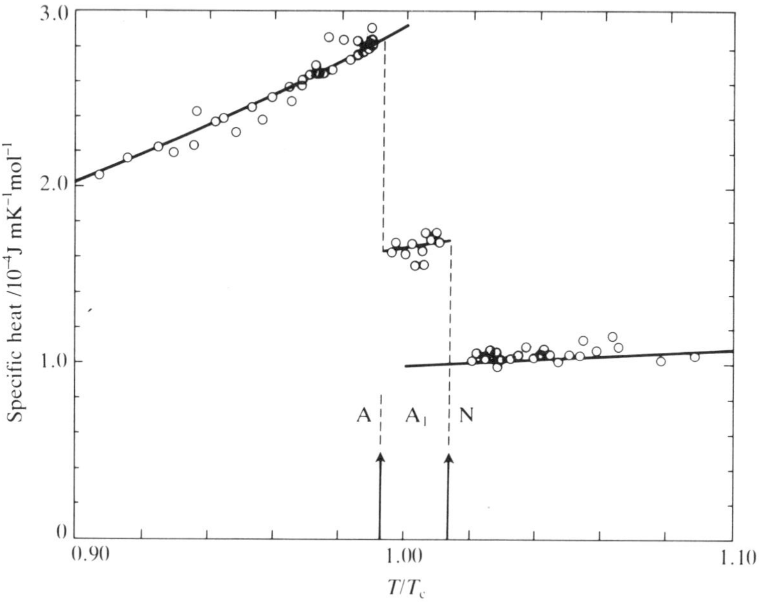

Fig. 9.15. The specific heat of liquid ^{3}He at the melting pressure in a magnetic field of 0.88 T. The data are shown in the vicinity of T_c where the A_1 phase could be observed. The solid lines in the A and N regions are fits to zero field data for ^{3}HeA and ^{3}HeN respectively. From Halperin *et al.* [99].

In the case of the B-phase, described by the BW model (see Section 9.7), we again have $L = S = 1$ but now the state function includes all three of the spin basis states as in eqn (9.2), and the total angular momentum quantum number is well-defined with the value $J = 0$. The state is isotropic in the sense that the gap Δ has the same value for all values of $\boldsymbol{k}$. However, the vector $\boldsymbol{d}$ is not the same for all values of $\boldsymbol{k}$ as it is in the ABM state but is a function $\boldsymbol{d}(\boldsymbol{k})$. This means that we cannot use either $\boldsymbol{l}$ or $\boldsymbol{d}$ as macroscopic vectors in ^{3}HeB although they still retain their microscopic meanings in relation to single pairs. Nevertheless the liquid exhibits a correlated structure, with a symmetry axis $\hat{\boldsymbol{n}}$, in which a pair associated with particular points $\pm\boldsymbol{k}$ on the Fermi surface has a value of $\boldsymbol{d}$ depending on the angle between $\boldsymbol{k}$ and $\hat{\boldsymbol{n}}$.

To describe this relationship let us fix one axis by applying a small magnetic field $\boldsymbol{B}$, and choose as a reference state one in which each pair (with $\pm\boldsymbol{k}$) has a $\boldsymbol{d}$-vector parallel to $\boldsymbol{k}$ (this reference state is not the equilibrium state, but is easier to visualize as a starting point). Now suppose the symmetry axis $\hat{\boldsymbol{n}}$ makes a random angle θ to $\boldsymbol{B}$ as shown in Fig. 9.16. Looking at that figure, imagine the tip of the vector $\boldsymbol{d}_i$ (i = initial) stuck to the surface of an imaginary cylinder whose axis is

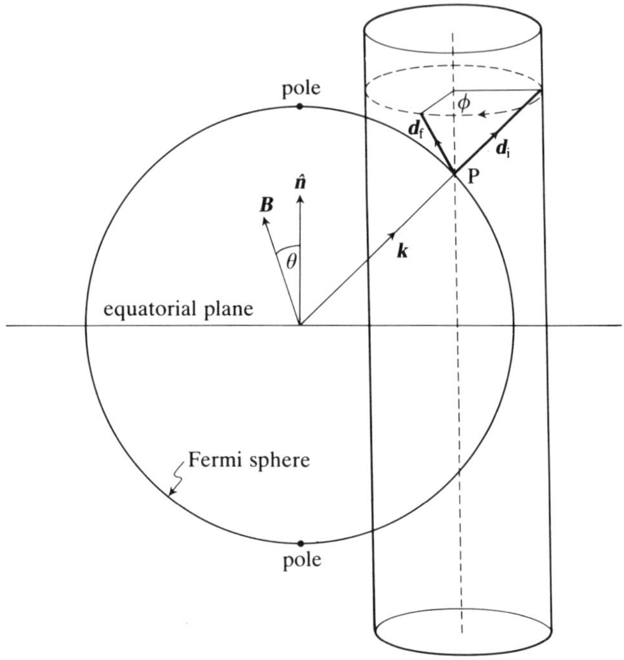

Fig. 9.16. A geometrical construction showing a rotation of the $\boldsymbol{d}$-vector for a pair with wave vector $\boldsymbol{k}$ represented by an arbitrary point P on the Fermi sphere. The initial reference position $\boldsymbol{d}_i$ lies parallel to $\boldsymbol{k}$, and the final position is shown as $\boldsymbol{d}_f$. For further explanation, see the text.

parallel to $\hat{\boldsymbol{n}}$, and $\boldsymbol{d}_i$ is parallel to the wave-vector $\boldsymbol{k}$ characterizing the pair represented by the point P on the Fermi sphere. Rotation of the cylinder about its axis parallel to $\hat{\boldsymbol{n}}$ by an angle ϕ takes $\boldsymbol{d}_i$ to $\boldsymbol{d}_f$ (f = final). A calculation of the dipole interaction energy between the nuclei of the pair shows that the energy is least when $\theta = 0$ and $\phi = \cos^{-1}(-\frac{1}{4}) = 104.48°$. That is, for a pair with P at a pole (see Fig. 9.16), $\boldsymbol{d}_f = \boldsymbol{d}_i$, both being parallel to $\boldsymbol{k}$, $\hat{\boldsymbol{n}}$ and $\boldsymbol{B}$. For a pair with P on the equator, the least energy occurs when $\boldsymbol{d}_f$ lies in the equatorial plane at an angle of 104.48° to $\boldsymbol{k}$ (also in the equatorial plane); in this case $\boldsymbol{d}_f$ is perpendicular to $\hat{\boldsymbol{n}}$ and $\boldsymbol{B}$.

We have seen that ^{3}HeB is convincingly superfluid [83], and some of its other properties have already been mentioned, but its spin dynamics are much more complicated that those of ^{3}HeA. Even so, equations of motion for $\hat{\boldsymbol{n}}$ and $\boldsymbol{S}$, equivalent to Leggett's eqns (9.3) and (9.4), have been obtained by Brinkman [101] and by Maki and Hu [102]. These equations have been solved using approximations to produce expressions analogous to eqns (9.5) and (9.6) for the behaviour of the system during both perpendicular and parallel ringing. Webb *et al.* [103] have observed parallel ringing in ^{3}HeB after turning off an incremental field $\boldsymbol{\Delta B}$ parallel to the steady field $\boldsymbol{B}_0$ but only at temperatures very close to T_C. Pulsed NMR experiments have also been made to investigate perpendicular ringing. If the tipping angle is less than $\cos^{-1}(-\frac{1}{4})$ there is a solution with zero frequency shift, but if the tipping angle is greater than $\cos^{-1}(-\frac{1}{4})$ then a shift is predicted, and has been observed by Osheroff and Corruccini [104].

It appears that textural effects play a role in a number of experiments on ^{3}HeB and they have been investigated by several authors. Osheroff *et al.* [105] measured perpendicular ringing shifts in narrow channels separated by flat plates, and deduced the orientation of $\hat{\boldsymbol{n}}$ with respect to $\boldsymbol{B}$ in this restricted geometry. In the bulk liquid $\hat{\boldsymbol{n}}$ and $\boldsymbol{B}$ would be parallel (see above) but the proximity of the surfaces alters this. An analysis of several terms in the free energy resulting in textural effects has been given by Brinkman *et al.* [106], and these include the bending of $\hat{\boldsymbol{n}}$, and the relation between $\hat{\boldsymbol{n}}$ and $\boldsymbol{B}$ at or near solid surfaces.

9.11 Acoustic experiments

Superfluid ^{4}HeII will propagate not only first (ordinary) sound but also second sound (temperature waves) third sound (surface waves) and fourth sound (waves in capillaries small enough to clamp the normal component). All these modes can be expected in the superfluid phases of ^{3}He, although not all have yet been observed. The experimental observation of fourth sound [87], see Section 9.5, supports the two-fluid model and implies that the other modes should also be observed. However the low value of the predicted velocity of second sound and the large value of the damping will present considerable experimental difficulties.

In addition to the modes associated with the two-fluid model, there are also zero sound and spin waves in superfluid ^{3}He as in ^{3}HeN. Furthermore there appear to be collective modes, which in some cases are coupled to longitudinal zero sound and contribute to the attenuation, measured by standard acoustic methods. These modes can be thought of as excited pair states, with energy quanta $\hbar\omega$ above the ground state, lying in the energy range between the pair ground state energy and the unpaired quasiparticle continuum. The energy of the modes may be characterized by the dimensionless ratio $\hbar\omega/\Delta$, where Δ is the gap energy, and if frequencies ω of order $\Delta/\hbar$ are used in the measurements then the pairs may be broken apart by quanta of zero sound, with a consequent increase in attenuation. However, by choosing a frequency high enough to generate zero sound, but not so high as to disrupt the pairs ($1/\tau \ll \omega \ll \Delta/\hbar$), it is possible to investigate pairs which are excited above their ground state but not broken.

In ^{3}HeA there is evidence for three collective modes named normal-flapping, clapping and super-flapping associated with different relative motions of $\boldsymbol{l}$ and $\boldsymbol{d}$ (see for example Dobbs [76]). These are listed in Table 9.2 below (from Wölfle [107]) together with values of the frequencies at which resonances occur, in terms of the energy gap $\Delta(T)$. It is difficult to form a satisfactory physical picture of these processes because the components of $\boldsymbol{d}$ must be treated as *complex*. Hence in a full treatment $\boldsymbol{d}$ must be expressed as *two* vectors which, together with $\boldsymbol{l}$, form a mutually perpendicular triad. The reader may imagine his vertical back to represent $\boldsymbol{l}$ and his arms, held horizontally and mutually perpendicular, to represent the other two vectors. Then 'clapping' corresponds to the arms oscillating in antiphase in the horizontal plane and 'flapping' to the arms oscillating in phase up and down.

Table 9.2. Resonant frequencies of collective modes in ^{3}HeA

Name	$\hbar\omega/\Delta(T)$
Normal-flapping	$(4/5)^{\frac{1}{2}}(T/T_c)$
Clapping	1.23
Super-flapping	1.56 (as $T \rightarrow 0$)
	2 (as $T \rightarrow T_c$)

There are also collective modes in ^{3}HeB which are characterized by the total angular momentum quantum number J, J_z, and a label i (imaginary) or r (real) which refers to the mathematical description. The notation for zero sound itself is $\{0, 0, \mathrm{i}\}$. Ketterson *et al.* [108] list twelve collective modes having $J = 0$, 1, and 2. The 'squashing' mode $\{2, 0, \mathrm{i}\}$ is strongly coupled to zero sound and gives the attenuation peak shown in Fig. 9.17. The other prominent peak in the figure, apart from that due to pair breaking, is due to the weakly coupled 'real squashing' mode $\{2, 0, \mathrm{r}\}$. In addition, when displayed on a much finer scale (see [109]), the data clearly show two much less prominent features at (T/T_c) values of 0.92 and 0.93 and it was at first speculated that these might correspond to further modes $\{0, 0, \mathrm{r}\}$ and $\{4, J_z, \mathrm{i}\}$ as suggested by the theoretical work of Sauls and Serene [111]. The latter mode, with $J = 4$, would have been particularly interesting as it would have indicated the presence of some pairs with $L = 3$ instead of the usual $L = 1$. However, further experimental work [112] showed that the features became much more pronounced

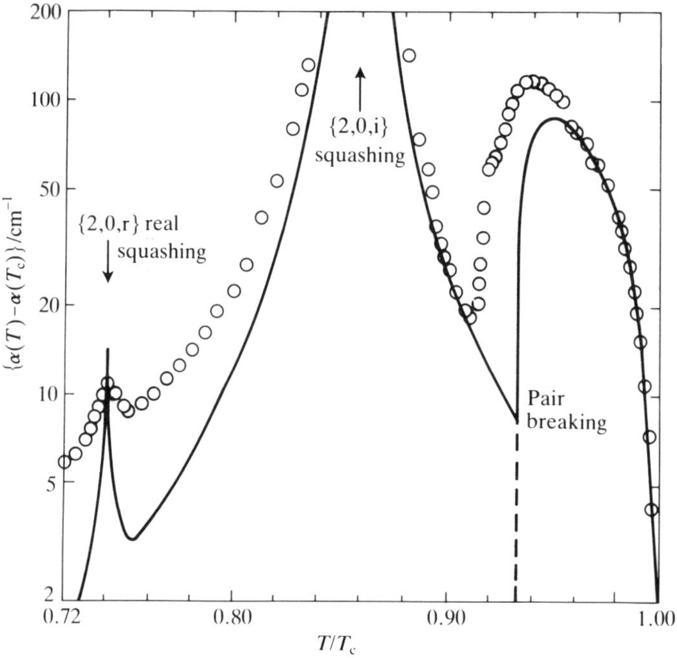

Fig. 9.17. The temperature dependence of the relative attenuation of 44.2 MHz longitudinal sound in ^{3}HeB at 2.44 b. The data are from Daniels *et al.* [109] and the continuous line represents a fitting to the theory of Wölfle [110].

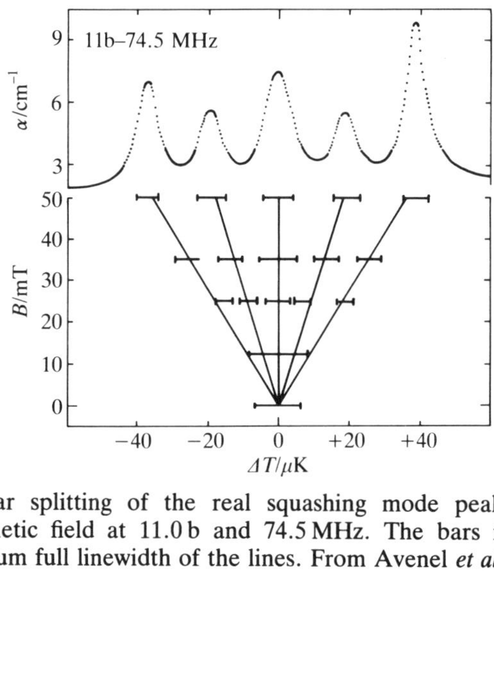

Fig. 9.18. Linear splitting of the real squashing mode peak in ^{3}HeB with transverse magnetic field at 11.0 b and 74.5 MHz. The bars indicate the half maximum full linewidth of the lines. From Avenel *et al.* [114].

when the magnetic field was raised, and this finally led to their identification by Schopohl and Tewordt [113] as $\{1, 1, \mathrm{i}\}$ and $\{1, -1, \mathrm{i}\}$.

The effect of magnetic fields on some absorption peaks is quite striking. In the case of the 'real squashing' mode, the application of a magnetic field transverse to the propagation direction splits the 'real squashing' mode peak into five components (Fig. 9.18). This Zeeman-type effect confirms the presence of the $J = 2$ state which is of course consistent with the model of p-wave pairing, with $S = 1$ and $L = 1$, which seems to be adequate for most purposes.

10

DILUTE SOLUTIONS OF ^{3}He IN HELIUM II

10.1 The phase diagram

Above 0.87 K liquid ^{3}He and ^{4}He are miscible in all proportions, but at lower temperatures a phase separation occurs as shown in Fig. 10.1. In this figure as elsewhere we denote the concentration by $X_3 = N_3/(N_3 + N_4)$ where N_3 and N_4 are the numbers of ^{3}He and ^{4}He atoms. X_3 is sometimes quoted as a fraction, sometimes as a percentage.

The lambda point of pure ^{4}He ($X_3 = 0$) is at 2.17 K but the addition of ^{3}He lowers T_λ until the tricritical point is reached at 0.87 K with $X_3 = 0.67$. The lambda line and the phase separation lines divide the phase diagram into three regions: (i) essentially normal liquid ^{3}He/^{4}He mixture; (ii) essentially superfluid liquid ^{4}He with some dissolved ^{3}He; and (iii) a two-phase region in which the liquid separates into ^{3}He rich and ^{4}He rich solutions. As the temperature is reduced towards absolute zero, the lighter phase consists of almost pure ^{3}He, with a ^{4}He concentration of [115]

$$X_4 \equiv (1 - X_3) = 0.85T^{\frac{3}{2}} \exp(-0.56/T) \tag{10.1}$$

while the heavier ^{4}He rich phase contains about 6 per cent of ^{3}He, or more precisely [116]

$$X_3 = 0.0648(1 + 8.4T^2 + 9.4T^3). \tag{10.2}$$

10.2 Dilute solutions of ^{3}He in liquid ^{4}HeII

The behaviour of dilute solutions of ^{3}He in liquid ^{4}He below about 0.5 K gives important information on the Fermi liquid. This comes about because the heat motion in liquid ^{4}He has then almost completely disappeared, its entropy and specific heat being very much less than that of liquid ^{3}He. Hence the liquid ^{4}He in the solution behaves only as a 'background fluid', in which the ^{3}He atoms move about rather as a gas of atoms in a vacuum. However, as a moving ^{3}He atom has to push aside ^{4}He atoms, the true inertial mass m_3 of the ^{3}He atoms must be replaced by a larger effective mass m_3^*, about 2.4 times greater than m_3.

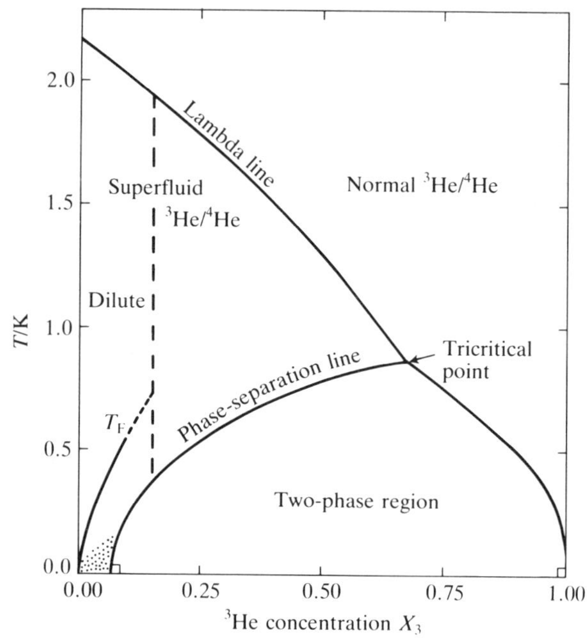

Fig. 10.1. The phase diagram of liquid ^{3}He/^{4}He mixtures with ^{3}He concentration X_3 plotted horizontally and temperature T vertically. See Sections 10.1 and 10.2 for explanations. (The figure has been constructed by taking data from Refs. [115–119]).

As we may largely disregard the presence of the ^{4}He atoms in a ^{3}He–^{4}He solution at low temperatures, we expect the ^{3}He atoms to be described by the Fermi liquid theory. Moreover we have the possibility of making experiments on a fluid ^{3}He of variable number density, and thus with different strengths of interactions between the atoms. In sufficiently dilute solutions the interactions will be very small, and the behaviour approximates to that of a perfect Fermi gas. This is the region of the phase diagram labelled dilute (with $X_3 \leq 0.15$) in Fig. 10.1. The figure also gives a line showing the degeneracy temperature T_f of the 'gas' of ^{3}He atoms as a function of concentration. The regions of classical and Fermi-degenerate behaviour lie on either side of this line, and the completely degenerate Fermi region is indicated by the dotted area in the lower left-hand corner.

The degeneracy temperature for the 'gas' of ^{3}He atoms is given by

$$T_f(X_3) = \frac{h^2}{8k_B m_3^*}\left[\frac{3NX_3}{V}\right]^{\frac{2}{3}}. \tag{10.3}$$

Numerically, $T_f = 2.58\ X_3^{\frac{2}{3}}\ K$; for example, in a 1 per cent mixture ($X_3 = 0.01$) $T_f = 120$ mK. The superfluid phases of pure ^{3}He, discussed in Chapter 9, occur very close to the right hand corner of the figure and no attempt is made to represent them. The question of possible superfluidity in the *dilute* ^{3}He 'gas' is also of interest. It had been hoped that under suitable conditions the transition might occur no lower than a few tenths of a millikelvin [120]. However it now seems [121] that the transition temperature may be experimentally inaccessible ($T_C \leqslant 0.05\ \mu$K). (We refer here to superfluidity in the *solute* ^{3}He; the *solvent* ^{4}HeII is certainly superfluid below T_λ.)

Figure 10.2 shows the specific heat of solutions, of various concentrations, between 0.1 and 0.5 K. At the higher temperatures the specific heat C approaches a value of about $\frac{3}{2}X_3R$ corresponding to the classical value of $\frac{3}{2}k_B$ per atom of ^{3}He. (The discontinuities for concentrations greater than $X_3 = 0.06$ are associated with the phase separation.) Note also that at temperatures much above 0.5 K, the specific heat of the solvent ^{4}He rises significantly and must be subtracted from the total before making comparisons with theory.

At lower temperatures, the specific heat falls below the classical value and eventually becomes linear in the temperature, as may be seen for 1.3

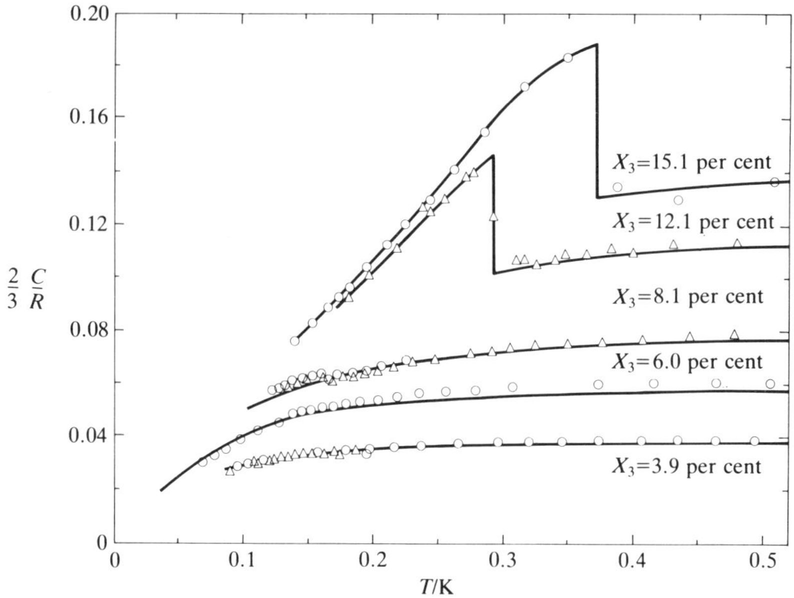

Fig. 10.2. The specific heat of liquid ^{3}He/^{4}He mixtures at the saturated vapour pressure plotted as $\frac{2}{3}C/R$ where R is the molar gas constant (from Edwards *et al.* [122]).

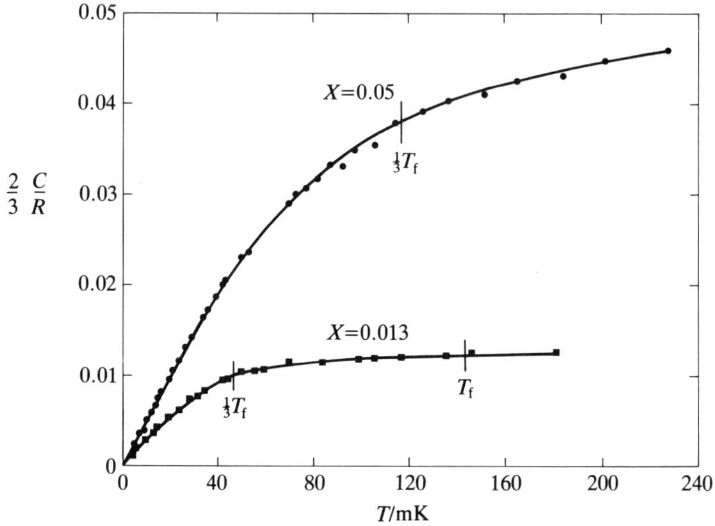

Fig. 10.3. The molar specific heat C of two solutions of ^{3}He in liquid ^{4}He at the saturated vapour pressure, plotted as $\frac{2}{3}C/R$ where R is the molar gas constant. (After Anderson *et al.* [123].)

and 5 per cent solutions in Fig. 10.3. At the lowest temperatures, the specific heat has the value we would expect if the effective mass of the ^{3}He atoms was equal to 2.4 m_3, where m_3 is the atomic mass, in fair agreement with estimates obtained in other ways. The full lines in the figure show the calculated values of the specific heat of a perfect Fermi gas of the same number density ($C/R = \frac{1}{2}\pi^2 X_3 T/T_f$) and we see that it coincides closely with the measured points. Finally we mention that the magnetic susceptibility of dilute solutions down to at least 0.35 K also exhibits the temperature dependence which we would calculate for a Fermi gas with an effective mass of about 2.4 m_3.

10.3 ^{3}He and the normal fluid

^{3}He solute atoms do not take part in superfluid flow. We have already mentioned in Section 5.6 that if helium II is allowed to run out of a vessel through a very fine capillary, only the superfluid passes through the capillary, and virtually all the normal fluid is left behind in the vessel. If we now repeat the experiment with a dilute solution in the vessel, we find that the ^{3}He atoms do not pass through the capillary, but remain behind with the normal fluid.

As ^{3}He solute atoms do not pass through a superleak, a fine capillary

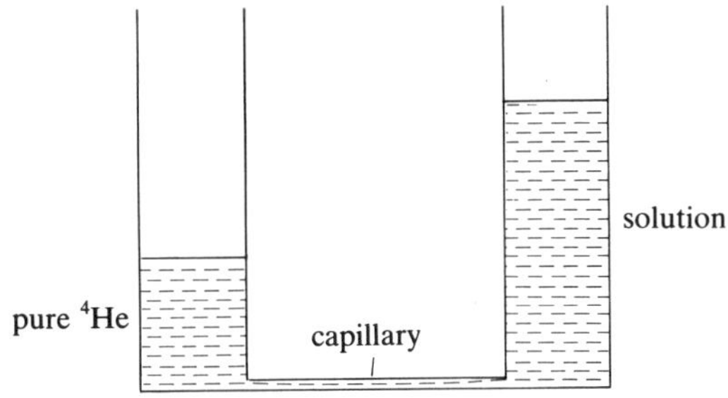

Fig. 10.4. Schematic diagram of apparatus for observing the osmotic pressure of ^{3}He in liquid ^{4}HeII.

may be used as a form of semi-permeable membrane to observe their osmotic pressure, as in Fig. 10.4. If both vessels are held at the same temperature to avoid any thermomechanical effects, superfluid from the left hand vessel passes through the capillary until the osmotic pressure of the ^{3}He is balanced by the excess hydrostatic pressure. The osmotic pressure may correspond to a large difference in levels, for example at the lowest temperatures a 1 per cent concentration of ^{3}He produces a difference of about 20 cm of helium, and is best measured either by using a pressure-sensitive diaphragm or by balancing the osmotic pressure against a fountain pressure. Figure 10.5 shows some results. In the

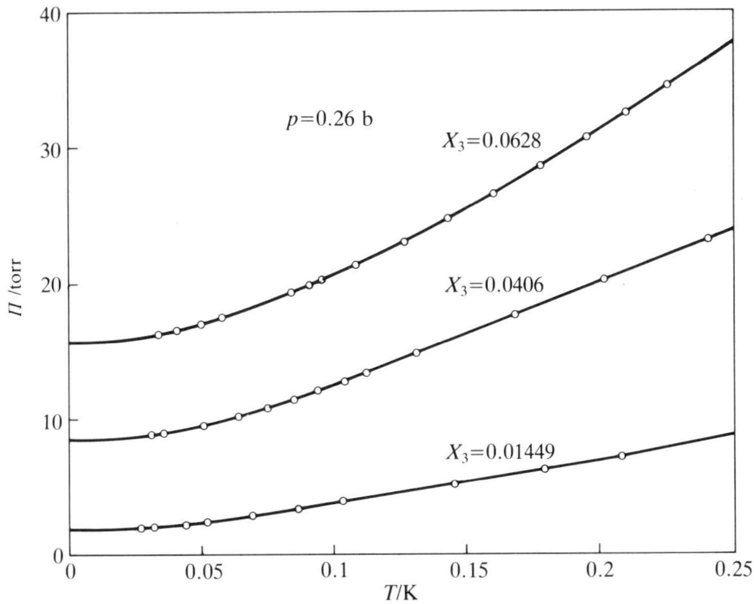

Fig. 10.5. Osmotic pressure of liquid mixtures as a function of temperature at three concentrations at a pressure of 0.26 atm. (From Landau *et al.* [124]).

classical limit ($T \geqslant T_f$), the osmotic pressure Π follows Van't Hoff's law (a direct analogue of the classical perfect gas law) within a few per cent as given by

$$\Pi = NX_3 k_B T/V.$$

In the degenerate limit ($T \ll T_f$) Π approaches a temperature independent value (this again is qualitatively like the behaviour of a Fermi degenerate gas whose pressure approaches the constant value $\frac{2}{5}N_3 k_B T_f(N_3/V)/V$, proportional to $(N_3/V)^{\frac{5}{3}}$). For more details see Ebner and Edwards [116]. Note that this account of the osmotic pressure implies that virtually no ^{3}He atoms pass through the capillary.

Another characteristic phenomenon associated with dilute solutions is the so-called 'heat flush' effect. If a heat current flows through a dilute solution, it is found that the ^{3}He concentration at the cold end of the system increases, while at the hot end it decreases. Indeed, if the heat flux is sufficiently large, virtually all the ^{3}He may be swept away from the hot end, hence the term 'heat flush'. (It is often necessary to beware of this effect when making measurements on dilute solutions!) From the point of view of the two-fluid model, it appears that the solute atoms are dragged along by the normal fluid as it moves from the hot to the cold parts of the system.

As we have just seen, the heat flush effect suggests that the ^{3}He atoms move with the normal fluid. This is confirmed by measurements of the normal density ρ_n of dilute solutions, using Andronikashvili's method of

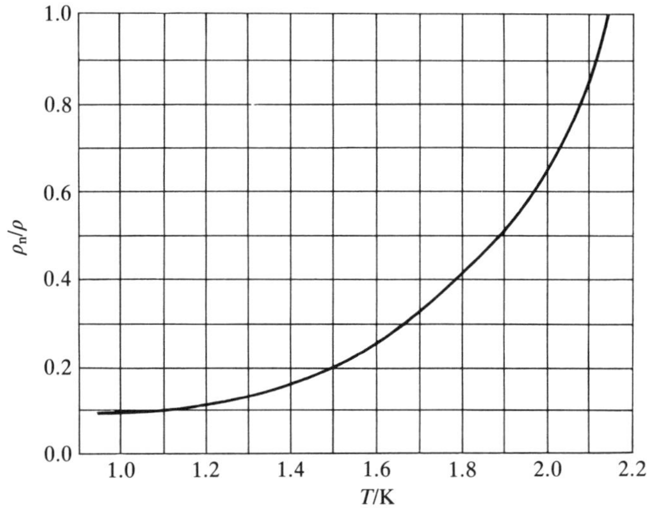

Fig. 10.6. The density of the normal fluid ρ_n in a 3.3 per cent (by mass) solution of ^{3}He in helium II, as a function of temperature (after Pellam [125]).

observing the period of a pile of discs oscillating in the liquid (Section 5.5). In pure helium II the normal density falls off below 1 K, as the heat energy tends to zero, but in solutions the ^{3}He atoms contribute to the normal fluid, so that the normal density tends to a constant and finite value (depending on the ^{3}He concentration) as shown in Fig. 10.6. Other experiments show that this additional contribution to ρ_n is directly proportional to the concentration of ^{3}He atoms and so it is possible to obtain in this way values of m_3^* as a function of X_3.

10.4 ^{3}He solute atoms as excitations

The experiments described in the previous section suggest that ^{3}He solute atoms behave rather as another form of excitation. They exert an osmotic pressure and cannot pass through superleaks. They contribute to the normal fluid, and are dragged along by a heat current. Like phonons and rotons, they will have an energy ϵ and a momentum p. The natural relation between these two quantities is the form originally proposed by Pomeranchuk, namely

$$\epsilon = \epsilon_{30} + p^2/2m_3^*. \tag{10.1}$$

The solutions are assumed to be sufficiently dilute that any interactions between the ^{3}He atoms are negligible, so that ϵ_{30} is the potential energy of a single ^{3}He atom in pure helium II. The second term on the right of (10.1) represents the kinetic energy of the atom. As is usual in hydrodynamics, the mass which appears in this term is not the true mass, but a somewhat larger effective mass m_3^* which takes account of the fact that a moving ^{3}He atom must push aside the ^{4}He atoms in its track. (We enlarge on this point in Section 11.5.) The character of these ^{3}He atoms is now completely specified by the parameters ϵ_{30} and m_3^*.

Small concentrations of ^{3}He make little difference to the properties of liquid ^{4}He above the lambda point. This is quite natural, for helium I behaves very much as a dense classical gas, and classically a ^{3}He atom is very similar to a ^{4}He atom. However, the position below the lambda point is quite different, and the liquid must be described in terms of elementary excitations. To anticipate the next chapter, the behaviour of helium II is the consequence of the Bose statistics and the symmetry of the wave function. All the ^{4}He atoms are indistinguishable from each other, and it is meaningless to talk of individual atoms. However, a ^{3}He solute atom is obviously distinguishable from a ^{4}He atom, and we therefore expect it to behave differently.

The essential characteristic of the ^{3}He solute atoms is that they are *different* from the ^{4}He atoms; the fact that they obey Fermi–Dirac rather than Bose statistics is irrelevant. This point is not too easy to demonstrate experimentally as the only substance soluble in helium II and obeying Bose statistics is the isotope ^{6}He, which has a half-life of only 0.8 s! Even so, it appears that this isotope does not take part in superfluid flow (Guttman and Arnold [126]).

10.5 Some dynamic properties

We now briefly indicate some of the principal ways in which ^{3}He solute atoms modify the dynamic behaviour of helium II. Figure 10.7 shows the effect of 1 per cent of ^{3}He on the normal viscosity η_n. The observed decrease comes about because the viscosity is proportional to the mean free path of the excitations, which is reduced by the ^{3}He atoms scattering the excitations. The effect is most marked at lower temperatures where the excitations are less numerous and produce less scattering themselves, so that the role of the ^{3}He is relatively more important.

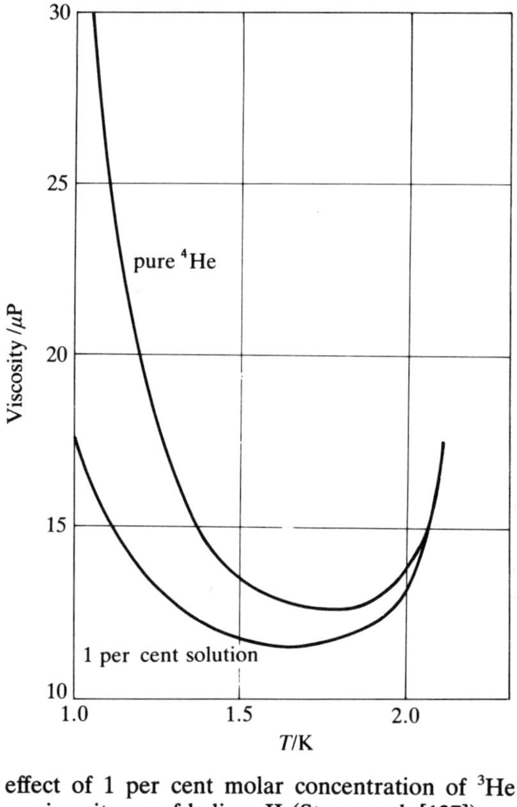

Fig. 10.7. The effect of 1 per cent molar concentration of ^{3}He on the normal viscosity η_n of helium II (Staas *et al.* [127]).

The effect of ^{3}He atoms on the heat flow is even more marked. The moving normal fluid sweeps the ^{3}He atoms to the cold part of the system, where they build up a concentration gradient tending to diffuse back to the warmer region. Scattering between these atoms and the normal fluid acts as a resistance to the flow of the normal fluid, which greatly reduces the heat flux. In fact for an 0.1 per cent solution the observed heat flow is only of the order of $5 \times 10^{-3}\,\mathrm{W\,cm^{-1}\,s^{-1}\,K^{-1}}$. Heat flow in solutions is affected by scattering processes involving phonons and rotons and by the heat flush effect which sweeps ^{3}He atoms along with the heat current to produce a concentration gradient in the liquid. However, at sufficiently low temperatures the numbers of phonons and rotons become negligible, and the heat is carried only by the 'gas' of ^{3}He atoms. If the temperature is also low enough for this 'gas' to be degenerate, then as in Section 3.2 we expect the mean free path to vary as $(T_f/T)^2$, and this leads to a crude prediction that the thermal conductivity should be proportional to X_3/T. Experiments show a fair agreement with this dependence on temperature (for example, measurements on a 1.3 per cent solution show $KT \approx 11\,\mathrm{erg\,s^{-1}\,cm^{-1}}$ [128], but more experimental evidence is needed on the dependence on concentration. For example, the curved line in Fig. 10.8 shows a recent theoretical prediction, whereas a simple degenerate Fermi gas model predicts a straight line through the origin; the present results are insufficient to resolve the difference. The simple treatment also predicts that the viscosity will be proportional to $(X_3)^{\frac{5}{3}}/T^2$, but this relationship may be modified by surface effects (see the recent work of Ritchie [129]).

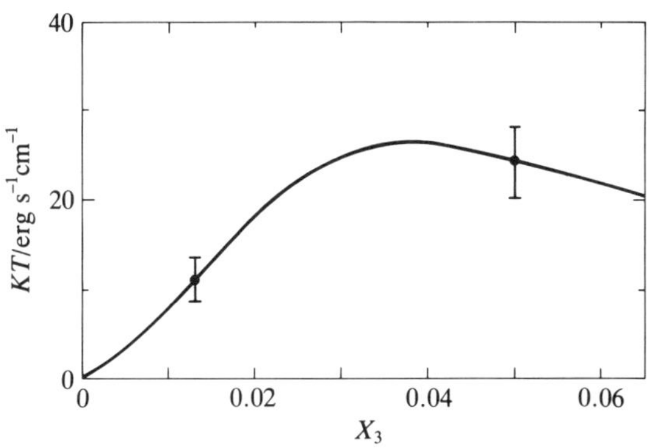

Fig. 10.8. The product KT of the thermal conductivity coefficient and the temperature, as a function of concentration X_3. The continuous line is from a theory [121] based on a choice of an assumed interaction potential between ^{3}He atoms, the choice being made in such a way as to optimize agreement with as many experiments on different properties of mixtures as possible. The two data points are from Abel *et al.* [128].

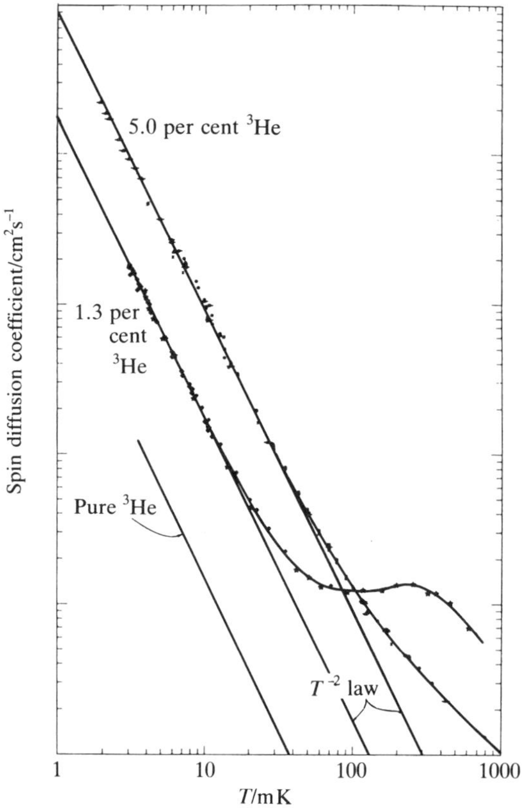

Fig. 10.9. The spin diffusion coefficient for dilute solutions of ^{3}He in liquid ^{4}HeII. (From Anderson *et al.* [123].)

The ^{3}He atom has a magnetic moment which gives the possibility of measuring another transport property, the spin diffusion coefficient, and one might expect D to vary as $(X_3)^{\frac{2}{3}}/T^2$. Figure 10.9 shows data for D; for $X = 0.013$ it is found that $DT^2 = 17 \times 10^{-6}\,\mathrm{cm^2\,K^2\,s^{-1}}$. More detailed theory can be found in Refs. [121] and [130] and a good earlier review of the subject has been given by Ebner and Edwards [116].

The scattering between the ^{3}He atoms and the excitations also shows itself in other ways. The large absorption of sound in the region of 1 K (Section 8.3) is considerably reduced by ^{3}He solute atoms, as shown in Fig. 10.10. A complete discussion of this effect is rather involved, but essentially the decrease arises because the additional scattering of the ^{3}He atoms permits equilibrium to be established more rapidly, and thus reduces the irreversibility associated with the changes of pressure.

We have already noted (Section 7.5) that the propagation of second

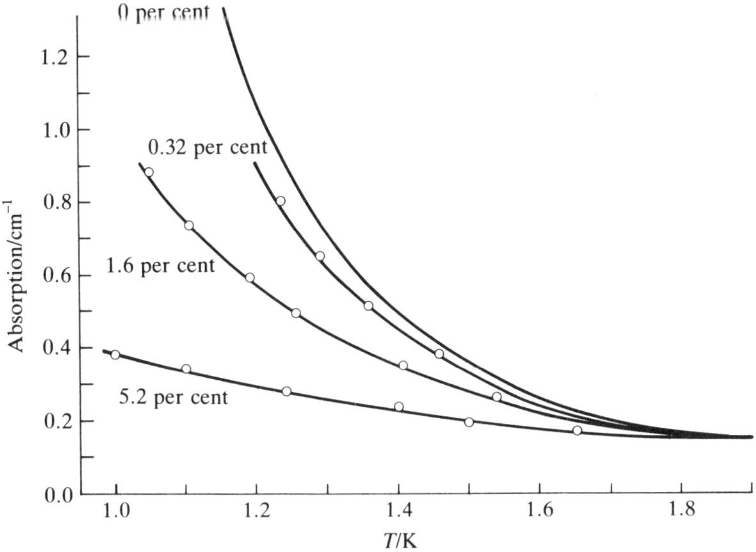

Fig. 10.10. The coefficient of absorption of 14.7 MHz sound waves in dilute solutions of ^{3}He in helium II. The figures on the curves indicate the molar concentrations of ^{3}He (Harding and Wilks [131]).

sound becomes impossible at some temperature below 1 K when the mean free path of the excitations becomes comparable with the wavelength of the sound. An addition of ^{3}He solute atoms reduces the mean free path, and therefore permits the observation of second sound to lower temperatures. Figure 10.11 shows the received signal when a 20 μs heat pulse is transmitted through pure helium II at 0.45 K, and through a 0.32 per cent solution at the same temperature. In the first case the shape of the pulse is characteristic of ordinary (Fourier) heat propagation, whereas the quite sharp pulse in the solution is characteristic of second sound. The velocity of second sound is also modified by ^{3}He solute atoms. This comes about partly because the normal density is greatly increased at the lower temperatures (Fig. 10.6), and also because the equations of motion of the liquid (Section 5.2) must be extended to

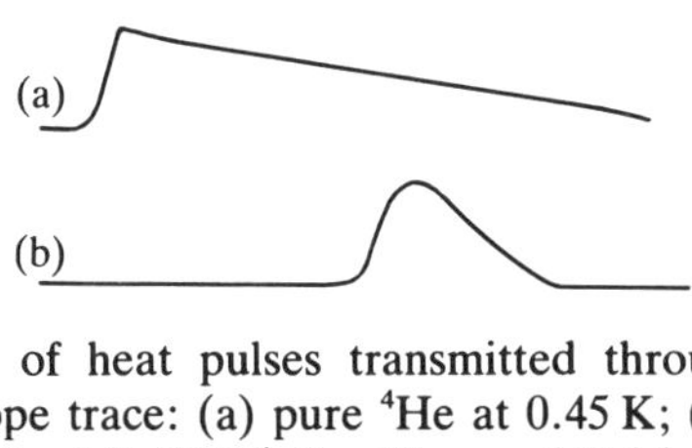

Fig. 10.11. The shape of heat pulses transmitted through liquid helium and viewed on an oscilloscope trace: (a) pure ^{4}He at 0.45 K; (b) ^{4}He containing 0.32 per cent ^{3}He at 0.45 K (after King and Fairbank [132]).

include two additional equations to specify the conservation of the ^{3}He atoms, and their effect on the chemical potential of the liquid.

10.6 Ions in helium II

We now refer briefly to some experiments involving the motion of positive and negative helium ions in liquid helium II, which can be produced in small concentrations by the action of suitable ionizing radiation. Their motion may then be directed by electric fields, and detected by measuring the resulting current. As both positive and negative ions are different from the uncharged ^{4}He atoms, we expect them to behave in rather the same way as ^{3}He solute atoms. In principle, a study of these ions should yield more information than experiments on ^{3}He solutions, because their motion is more easily observed. However, the theoretical treatment is more complicated because of the structure of the ions, see for example Schwarz [133].

The negative ion appears to be a bubble with a radius of about 1 or 2 nm (depending on pressure), inside which there is a single electron. The size is determined by a balance of surface tension and the pressure associated with the electron's zero-point energy. The positive ion is quite different: the central ion of $^3He^+$ attracts a few neighbours (by an essentially electrostrictive effect), forming a grain of solid helium of radius about 0.6 nm. Both types of ion appear to be approximately spherical and appreciably larger than the average interatomic spacing (0.35 nm at the vapour pressure as $T \to 0$).

When ions move under the influence of an electric field through normal helium (^{4}HeI or ^{3}HeN) at speeds corresponding to low Reynolds numbers, it is natural to think of their velocity being limited by an ordinary viscous drag. The ion accelerates in a field E to a steady drift velocity v_d, and the ratio $\mu = v_d/E$ is defined as the mobility. For a small sphere of radius r and charge q obeying Stokes' law in a fluid of viscosity η, the mobility would be given by $\mu = q/6\pi r\eta$. Values of the ion radii deduced in this way from measured mobilities in ^{4}HeI and ^{3}HeN are not strongly dependent on temperature. However in ^{4}HeII below the lambda point, rotons, phonons and any ^{3}He impurities scatter individually off the ions. Measurements of the mobilities of the ions at temperatures below about 1.3 K where the excitations themselves are quite well defined have led to a better knowledge of the ion structures, and have provided much detailed information about how they scatter the elementary excitations [133].

With high electric fields the observed behaviour changes radically because the ions create quantized vortex rings to which they then become attached (see Section 12.5). However, the creation of vortex rings by negative ions in ^{4}HeII can be effectively inhibited by rigorously excluding ^{3}He impurities and by maintaining pressures somewhat in excess of the vapour pressure. The reasons for using this recipe are discussed in detail by McClintock and various co-workers [134] and by Muirhead and Vinen [135]. The technique has been beautifully exploited to demonstrate that at low temperatures negative ions can indeed be accelerated without vortex creation to the Landau velocity, at which dissipation appears due to the creation of roton pairs. The Landau velocity, measured in this way, is shown in Fig. 10.12 as a function of pressure. The agreement between experiment (circles) and theory (continuous lines) is good, although it must be remembered that the theory is sensitive to the exact values of Δ, p_0 and μ characterizing the roton part of the dispersion

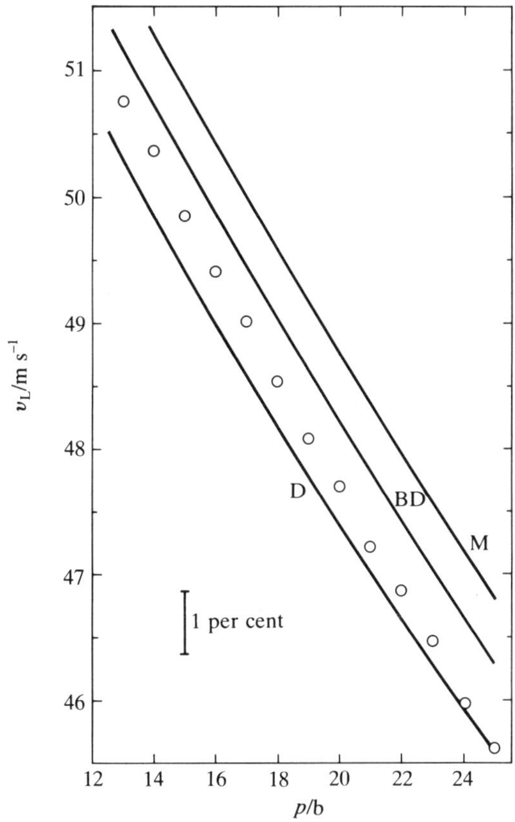

Fig. 10.12. The Landau critical velocity for roton creation v_L as a function of pressure p. The circled points are data and the continuous lines represent three different predictions depending on choices of the parameters Δ, p_0 and μ. From Ellis and McClintock [134].

curve, see eqn (6.3). The result is particularly satisfying because most other types of measurement give much lower critical velocities associated with the generation of vortex rings or the maintenance of vortex tangles as discussed in Chapter 14.

The ion technique has also been used to study the motion of small concentrations of ions in a heat current flowing in helium II. The ions were acted on by electric fields transverse to the direction of the heat flow. The magnitude and direction of their velocities were then deduced by observing the deflection of the ion currents flowing to a number of separate collector electrodes. Hence, it could be demonstrated that in the absence of a field, the ions move with the velocity of the normal fluid, as we have already assumed for ^{3}He solute atoms. We refer to other experiments using ions in Chapters 12 and 14.

10.7 The dilution refrigerator

The usual method of maintaining a low temperature is by the evaporation of low boiling point liquids. Thus a temperature bath between 4 and 1 K is readily obtained by pumping on liquid ^{4}He in a dewar vessel. To maintain the same cooling capacity at a lower temperature, we must evaporate the same amount of liquid at a lower pressure by using a larger pumping system. However, the saturated vapour pressure decreases very rapidly with temperature, so we soon reach a practical limit to the size of the pumping system, and to the lowest temperature obtainable. In the case of liquid ^{4}He this is about 1 K, but ^{3}He has a higher vapour pressure on account of the zero-point energy, so temperatures down to about 0.3 K are quite possible. As ^{3}He is only obtainable in relatively small quantities, it is usual to operate a ^{3}He bath within a larger cryostat maintained at about 1 K by a much larger quantity of ^{4}He. Details of such ^{3}He cryostats are given in Ref. [5]. Very low temperatures have been achieved since the early 1930s by the techniques of adiabatic demagnetization, and today that is still the only way of going below 1 mK, but so-called dilution refrigerators can now maintain temperatures down to about 4 mK. The arrangement of components in such a refrigerator is shown very diagrammatically in Fig. 10.13 and we describe its operation only briefly here (see Ref. [5] for more details).

A mixing chamber contains a phase-separated solution with almost pure liquid ^{3}He floating on a 6 per cent solution of ^{3}He in liquid ^{4}HeII. When the refrigerator is operating, ^{3}He is moved first to the upper phase of almost pure ^{3}He, then to the lower ^{4}He rich phase, from which it is distilled, and then finally returned by a different route to the ^{3}He rich phase. The cooling is produced in the mixing chamber by the expansion of the 'gas' of ^{3}He atoms as they pass from the ^{3}He rich phase into the

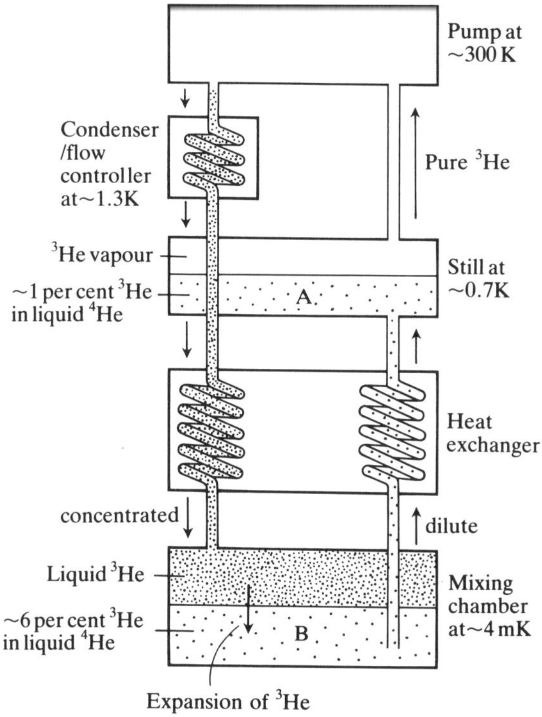

Fig. 10.13. The arrangement of components in a dilution refrigerator, illustrated very diagrammatically. See the text for an explanation of the principles of operation.

dilute condition of the ^{4}He rich phase. The heat extraction rate depends on the molar rate of ^{3}He flow across the interface in the mixing chamber, $\dot{n}_3$, and on the enthalpy difference between the condensed and expanded states of the ^{3}He. As both these states are Fermi-degenerate, the enthalpy difference is proportional to T^2, and numerically the heat extraction rate is $94\dot{n}_3T^2$ W. The magnitude of $\dot{n}_3$ is determined by the operating conditions and the speed of the pumping system, and is typically (within an order of magnitude) about 10^{-4} mol s^{-1}. The circulation of ^{3}He is maintained by distilling almost pure ^{3}He from the 1 per cent solution in the 'still' (Fig. 10.13) at about 0.7 K, and in order to maintain a given rate $\dot{n}_3$ it is necessary to supply heat to the still at a rate of about $40\,\dot{n}_3$ W. The ^{3}He lost from the still by distillation is steadily replaced by the passage of ^{3}He atoms across the phase interface in the mixing chamber and then upwards through the heat exchangers to the still. The pumped ^{3}He is condensed on the high pressure side of the pump and passes down through the heat exchangers where it is cooled before arriving in the mixing chamber, completing a cycle.

The lowest temperature for a particular refrigerator is reached when the heat extraction rate is just balanced by stray heat leaks and by any inefficiency in the heat exchangers. In fact the correct design of these components is crucial to the successful operation of the refrigerator. Many square metres of contact area are needed to overcome the Kapitza boundary resistance (values of which are given in Figs. 8.6 and 8.7), and the exchangers are usually produced by sintering fine metallic powders (often silver) to make porous blocks through which the helium can pass.

11
WAVE FUNCTIONS OF HELIUM II

We have already related many of the properties of helium II to the form of the excitation spectrum. We now outline Feynman's demonstration of why the excitation spectrum takes its characteristic form (Feynman [136, 137]).

11.1 The ground state of helium II and phonons

The liquid has a very open structure, in which the atoms may pass easily from one configuration to another. Even in the ground state at absolute zero, the atoms are not at rest, for they have considerable kinetic energy arising from the zero-point motion (Section 1.1). The strong repulsive forces ensure that atoms do not overlap, while the attractive forces have much less influence. In fact, we may think of the atoms as a collection of hard spheres, able to move about freely, and with a tendency to take up a fairly uniform distribution.

The wave function ϕ of the ground state will be a definite though complicated function of the coordinates $\boldsymbol{R}_i$ of the N atoms. The value of $|\phi(\boldsymbol{R}_1, \boldsymbol{R}_2, \ldots, \boldsymbol{R}_N)|^2$ indicates the probability of finding the atoms in the ground state in the particular configuration $\boldsymbol{R}_1, \boldsymbol{R}_2, \ldots, \boldsymbol{R}_N$. As described above, the atoms are free to move about in the liquid subject only to the condition that they must not overlap. Hence the value of ϕ must tend to zero for configurations in which atoms approach too closely. Moreover, just as the wave function for a particle in a box has its maximum value when the particle is at the centre of the box, and furthest from the walls, we expect the maximum value of ϕ to be for configurations in which the atoms keep well apart. In simple systems for which the Schrödinger equation has been solved completely, it turns out that the eigenfunctions have real values. We therefore assume that ϕ is real. In addition, ground-state wave functions have no nodes, hence we assume ϕ to have a positive value for all configurations of the atoms.

Feynman now discusses the wave functions of the excited states in a manner which gives nearly all the information we require about them, without having to calculate the complicated ground-state function. He begins by considering the low energy excitations which we know to be phonons. The equations of motion of sound vibrations expressed in normal coordinates are analogous to those of the harmonic oscillator. The wave function of the ground state of a linear harmonic oscillator with position coordinate x is a Gaussian function $\exp(-\frac{1}{2}\alpha x^2)$. The coordinate x fluctuates on account of zero-point motion, and the probability of a fluctuation of given amplitude falls off as the exponential of the square of the amplitude. The wave function of the first excited state is $x\exp(-\frac{1}{2}\alpha x^2)$, which is just the ground-state function multiplied by the position coordinate. Feynman recalls that the classical normal coordinate appropriate to sound waves in a liquid may be written

$$\rho_{\mathbf{k}} = \int \rho(\boldsymbol{R}) \exp(\mathrm{i}\boldsymbol{k}\cdot\boldsymbol{R})\, \mathrm{d}^3\boldsymbol{R},$$

where $\rho(\boldsymbol{R})$ is the number density, or alternatively, if we replace $\rho(\boldsymbol{R})$ by a quantum mechanical density operator,

$$\rho_{\mathbf{k}} = \sum_i \exp(\mathrm{i}\boldsymbol{k}\cdot\boldsymbol{R}_i).$$

(The first term in the bracket has the usual value $(-1)^{\frac{1}{2}}$, and should not be confused with the subscripts associated with the summation.) By analogy with the harmonic oscillator, Feynman now writes the excited state wave function as the ground-state function ϕ multiplied by ρ_{k}, that is

$$\psi_{\mathrm{ph}} = \left\{\sum_i \exp(\mathrm{i}\boldsymbol{k}\cdot\boldsymbol{R}_i)\right\}\phi. \tag{11.1}$$

It is not immediately obvious that the wave function (11.1) represents a density variation. However, we recall that the ground-state function corresponds to a great range of possibilities for the configurations of the atoms, some being more probable than others. Consider the real part of (11.1),

$$\left\{\sum_i \cos(\boldsymbol{k}\cdot\boldsymbol{R}_i)\right\}\phi.$$

It might seem that the cosine term will be positive for exactly half the number of configurations (sets of values $\boldsymbol{R}_1, \boldsymbol{R}_2, \ldots, \boldsymbol{R}_N$), and negative for the other half. In this case the function would vanish. However, the zero-point motion of the ground state can be regarded as an assembly of density fluctuations. The term $\sum_i \cos(\boldsymbol{k}\cdot\boldsymbol{R}_i)$ therefore picks out from the ground state just those fluctuations corresponding to phonons of wave number $\boldsymbol{k}$.

11.2 The scarcity of excited states

We now consider what other forms of excitations besides phonons are permitted in the liquid. As any density fluctuations may be described in terms of phonons, any other excitations must involve motions of the atoms at approximately constant density. As already noted, a wide range of configurations is available to the atoms, because of the very open structure of the liquid. Feynman therefore assumes the atoms can readily be moved from one configuration to another by some type of stirring process. To transfer the system from the ground state to an excited state involves rearranging the atoms into some new configuration. Our problem is to deduce the wave function of this configuration.

According to the general properties of the Schrödinger equation, the wave function of a stationary excited state ψ is orthogonal to the ground-state function ϕ. As ϕ is everywhere positive, ψ must be positive for some configurations $(\boldsymbol{R}_1, \boldsymbol{R}_2, \ldots, \boldsymbol{R}_N)$ and negative for others. Thus, if ψ is plotted against the $3N$ coordinates of configuration space, it will vary between certain maximum positive and negative values. We are concerned in the first place with states of low energy, and the criterion for such states is that the gradient of ψ in configuration space should be small. (The wave function of a particle in a box oscillates more frequently the higher the energy, as the kinetic energy is given by $-(\hbar^2/2m)\nabla^2\psi$.)

Let us refer to the configuration which gives ψ its maximum positive value as an A configuration, and that which gives the maximum negative value as a B configuration. Then if ψ is to represent a state of low energy, the points corresponding to configurations A and B should be widely separated in configuration space. If the atoms are distinguishable, this condition implies that to move atoms from an A configuration to a B configuration involves a stirring process in which all the atoms are displaced a long way. However, as all the atoms of the liquid are indistinguishable and obey Bose statistics, the wave function must remain unchanged by any permutation of the atoms. Hence it is possible to go from an A configuration to a B, or indeed to any other one, by moving each atom by a distance of no more than about the atomic spacing (see Fig. 11.1). This means that configurations A and B can only be separated by a certain limited distance in configuration space, the energy cannot be made arbitrarily small, and will not drop below a certain minimum value.

The above argument shows that some minimum energy is required to create excitations other than phonons. In particular we have ruled out the possibility of particle-like excitations with a spectrum $\epsilon = p^2/2m$. It also follows that the number of non-phonon excitations present in the liquid will decrease exponentially as the temperature is reduced. Hence at sufficiently low temperatures only phonons will be present. Note that the

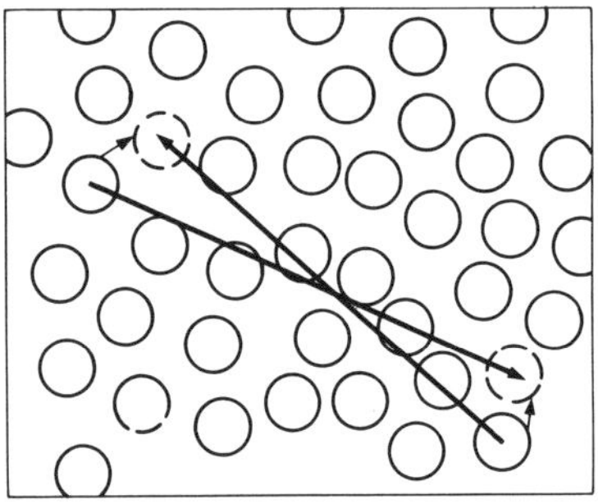

Fig. 11.1. The effect of two large displacements of the atoms (long arrows), can actually be accomplished by much smaller adjustments (short arrows) because of the identity of the atoms (Feynman [137]).

above conclusions will not apply to liquid ^{3}He, which obeys Fermi–Dirac statistics. The wave function then changes sign if we permute atoms, and it is very difficult to see what happens using this approach. The behaviour of ^{4}He is a consequence of the symmetry of the wave functions of a Bose system.

11.3 Feynman's variational method

In this section we describe Feynman's method for deducing the form of the excitation spectrum. This is based on the variational method. If a system has a Hamiltonian H, and ψ is any wave function orthogonal to the ground state, then the energy of the first excited state

$$E' \leqslant \frac{\int \psi^* H \psi \, \mathrm{d}\tau}{\int \psi^* \psi \, \mathrm{d}\tau}. \tag{11.2}$$

The equality sign is valid only if ψ is an exact eigenfunction of H. As a completely general expression eqn (11.2) is not particularly useful. Suppose, however, that we have a system for which the Schrödinger equation is very complicated, but for which we think we can guess an approximate wave function ψ. Then, if we guess well, the upper limit for E' given by (11.2) may be quite close to the true value. We may go further by expressing our trial wave function ψ in terms of some parameter or parameters which are of physical significance for the problem in question. In this case, if we vary the parameters so as to minimize the right-hand side of (11.2), we may expect the function ψ to become a better approximation to the true solution.

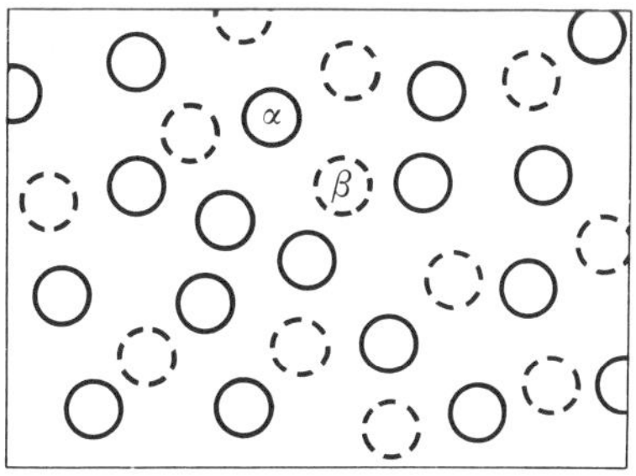

Fig. 11.2. The excited state wave function is positive for the configuration shown by solid circles (α-sites), and negative for the dotted circles (β-sites). For the lowest energy, all the β-sites must be as far as possible from the α-sites (Feynman [137]).

Feynman begins by choosing a trial wave function to describe an excitation in the liquid. Figure 11.2 shows schematically the sites of the atoms (labelled α and β) corresponding to the A and B configurations which give ψ its maximum positive and negative values (cf. Section 11.2). In order to obtain the states of lowest energy, an α-site must be as far away as possible from a β-site. Yet because of the Bose statistics, no α-site can be more than an atomic spacing from the corresponding β-site, hence the A and B configurations are quite similar. They will be farthest from each other in configuration space if it is necessary to move every atom to effect a transition between the two. That is, if all the α-sites are between β-sites, as shown in the figure.

The states of the system with lowest energy will be those in which the wave function varies smoothly as the atoms pass from an A to B configuration. To ensure this, the amplitude of ψ is taken to be the number of atoms on α-sites minus the number of atoms on β-sites. This number is $\sum_i f(\boldsymbol{R}_i)$ where $f(\boldsymbol{R})$ is a function which is $+1$ if $\boldsymbol{R}$ defines an α-site, and -1 if $\boldsymbol{R}$ defines a β-site, and which varies smoothly between these limits. We have also the condition that the atoms must not overlap, so, bearing in mind our treatment of the phonons, we choose as the trial function

$$\psi = \left\{\sum_i f(\boldsymbol{R}_i)\right\}\phi, \tag{11.3}$$

where ϕ is again the ground-state function, which takes account of correlations between the atoms. Although Feynman does not specify the motion of the atoms more precisely than by this last equation, he points out that (11.3) is consistent with three quite probable modes of motion, namely the rotation of a small ring of atoms, the excitation of single

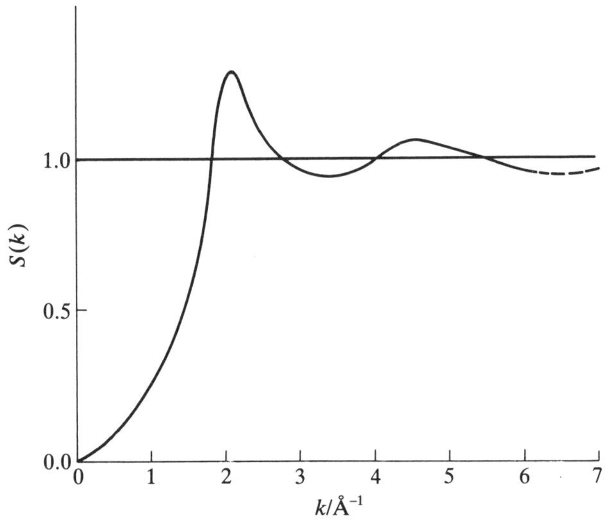

Fig. 11.3. The liquid structure factor $S(k)$ (Feynman and Cohen [138]).

atoms in a cage, and single atoms vibrating while the surrounding atoms move about to get out of their path.

The variational method is now used to determine the best form of the function $\sum_i f(\boldsymbol{R}_i)$. The Hamiltonian of the liquid is taken to be

$$H = -(\hbar^2/2m)\sum_i \nabla_i^2 + V - E_\phi,$$

where V is the potential energy of the system, and E_ϕ the energy of the ground state. After a rather long calculation, the ground state ϕ drops out, and we find that

$$f(\boldsymbol{R}) = \exp \mathrm{i}(\boldsymbol{k} \cdot \boldsymbol{R}), \tag{11.4}$$

which leads to an energy

$$E(k) = \hbar^2 k^2/2mS(k) \tag{11.5}$$

where $S(k)$ is the Fourier transform of the two-atom correlation function $p(\boldsymbol{R})$. This latter function is related to the probability of finding an atom at a certain distance from a given atom, and may be determined from X-ray scattering measurements. The form of $S(k)$ is shown in Fig. 11.3.

11.4 Feynman's first wave function

Substituting values of $S(k)$ into the expression for the excitation spectrum (11.5), we obtain curve A of Fig. 11.4. The curve reproduces qualita-

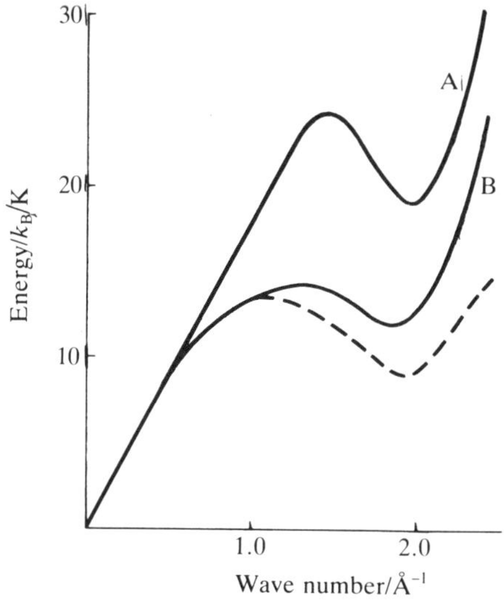

Fig. 11.4. The energy spectrum of excitations in helium II. Curve A as calculated by Feynmann [136]; curve B as calculated by Feynman and Cohen; dashed line as determined by neutron scattering experiments. (After Feynman and Cohen [138].)

tively the main feature of the results from the neutron scattering experiments, but the heights of the maximum and minimum are much too high.

The variational method satisfactorily reproduces the phonon spectrum for the lower values of k. Yet it is disconcerting that the wave function obtained by substituting (11.4) into (11.5), namely

$$\psi = \left\{\sum_i \exp(\mathrm{i}\boldsymbol{k}\cdot\boldsymbol{R}_i)\right\}\phi, \tag{11.6}$$

is for *all* values of $\boldsymbol{k}$ identical with the function we deduced for phonons in Section 11.1. Clearly the excitations of higher wave number are not phonons, because the dispersion relation is quite different. This discrepancy, like that in Fig. 11.4, shows that the wave function (11.6) is not a good approximation at high values of $\boldsymbol{k}$.

We note that the excitations specified by (11.6) are not localized, but have the form of plane waves passing through the liquid. However, they may be localized, just as the phonons in Section 6.1, by making up wave packets with linear combinations of wave functions.

11.5 ^{3}He solutions and the two-fluid model

At this stage it is instructive to refer to some of Feynman's calculations on the behaviour of a ^{3}He atom dissolved in helium II at absolute zero. As we saw in Chapter 10, a ^{3}He solute atom behaves in many ways as a particular form of excitation. However, for reasons which become apparent as we proceed, it is much easier to form a physical picture of the motion of a ^{3}He atom than of an excitation like a roton.

As the ^{3}He atom is distinguishable from all the ^{4}He atoms, it is not subject to the Bose statistics, there are few restrictions on its motion, and it has an energy spectrum of the normal classical form

$$\epsilon = p^2/2m_3^*. \tag{11.7}$$

However, the mass m_3^* is not the actual, but the effective, mass of the atom. As the ^{3}He atom passes through the helium II it displaces ^{4}He atoms which must move out of its way to let it pass. Therefore an acceleration of the ^{3}He atom involves the acceleration of nearby ^{4}He atoms. Hence, the apparent mass of the ^{3}He atom will be greater than its actual mass, as was shown by the experiments described in Chapter 10. This effect follows from classical hydrodynamics. A sphere moving through a perfect irrotational fluid experiences no frictional force whatever, but its apparent mass is augmented by a certain factor (which for a sphere in a liquid of the same density has the value 1.5).

Consider now a collection of small spheres (or ^{3}He atoms) moving in an irrotational fluid (or helium II at absolute zero). The ^{3}He atoms will collide with each other and with the walls of the container, and in accord with (11.7) will exchange energy and momentum as would free particles. There is no frictional force with the helium II, hence the ^{3}He atoms will appear to behave as a gas of particles of energy $\epsilon = p^2/2m_3^*$ moving freely in a vacuum. Thus the atoms form the 'normal fluid' associated with a solution at absolute zero. In addition there are other motions in the liquid which carry energy and momentum, namely the motions of the helium II as it flows out of the way of the ^{3}He atoms. This flow is irrotational and does not give rise to any viscous effects. Thus the total motion of the solution falls naturally into the two divisions of the two-fluid model.

The wave function for a ^{3}He atom moving freely through the liquid without any backflow would be

$$\psi = \{\exp(\mathrm{i}\boldsymbol{k}\cdot\boldsymbol{R}_\mathrm{A})\}\phi,$$

where $\boldsymbol{R}_\mathrm{A}$ is the coordinate of the atom. We may readily verify that this function leads to an energy $\hbar^2k^2/2m_3$, where m_3 is the actual mass of the atom. To take account of the backflow Feynman assumes another term in the wave function so that

$$\psi = \{\exp(\mathrm{i}\boldsymbol{k}\cdot\boldsymbol{R}_{\mathrm{A}})\}\left\{\exp \mathrm{i}\sum_{i\neq A} s(\boldsymbol{R}_i - \boldsymbol{R}_{\mathrm{A}})\right\}\phi, \tag{11.8}$$

this expression being chosen so that the motion is irrotational and has a physically realistic form (cf. Section 11.7).

Substituting (11.8) into the variational principle, we find that the requirement of minimum energy leads to a differential equation for $s(\boldsymbol{R})$ which for large values of $\boldsymbol{R}$ has the solution

$$s(\boldsymbol{R}) = A\boldsymbol{k}\cdot\boldsymbol{R}/R^3, \tag{11.9}$$

where A is some constant. We may now hope to obtain the lowest value of the energy by substituting for $s(\boldsymbol{R})$ in (11.8) and determining the value of A by carrying out another variation. After some heavy calculations it is found that

$$E = 0.648\hbar^2k^2/2m_3,$$

which corresponds to an effective mass of $1.55m_3$. This value for the energy is still too high, the effective mass determined experimentally being about $2.5m_3$. This discrepancy is not altogether surprising because, quite apart from the correctness of the wave function, several approximations are involved in the calculations. Thus we require the correlation functions for the ^{4}He atoms around the ^{3}He atom but this information is not available. Feynman and Cohen therefore assume that these functions are the same as for ^{4}He atoms around another ^{4}He atom, but this is obviously not exact.

11.6 Feynman and Cohen's wave function

The description of a ^{3}He solute atom in the previous section suggests a possible form of wave function for a roton. Let us picture a ^{3}He atom moving through helium II with an energy of about 9 K, the motion being accompanied by an irrotational backflow. If we now replace the ^{3}He atom by a ^{4}He atom we obtain some sort of excitation probably not very different from a roton. However, as we can no longer distinguish between any of the atoms, we must use a purely quantum mechanical description.

Any system of Bose particles must be described by a symmetrical wave function. Therefore Feynman and Cohen [138] suggest that the wave function for the situation described in the previous paragraph is obtained by symmetrizing (11.8) and writing

$$\psi = \left\{\sum_i \exp(\mathrm{i}\boldsymbol{k}\cdot\boldsymbol{R}_i)\right\}\left\{\exp \mathrm{i}\sum_{j\neq i} s(\boldsymbol{R}_i - \boldsymbol{R}_j)\right\}\phi. \tag{11.10}$$

The energy spectrum is obtained by inserting (11.9) and (11.10) into the variational equations, and finding the best value for A. After some computation the second excitation curve shown in Fig. 11.4 is obtained. It is clearly a big improvement over the earlier spectrum (11.5), but still gives too high an energy at the roton minimum. (Thus the calculated value of Δ/k_B is 11.5 K compared with the value of 8.67 K from the neutron scattering experiments.) However, taking into account the complexity of the calculations, the agreement must be regarded as fairly satisfactory.

Phonons are associated with the vibrational motions of the atoms, but no such simple picture seems possible for the rotons. We may visualize a ^{3}He solute atom moving through helium II, but this simple physical picture vanishes when we symmetrize the wave function to describe pure ^{4}He. However, the discussion of the last two sections suggests that a roton is essentially one ^{4}He atom travelling rapidly through the other atoms which move out of its path.

11.7 Superfluid flow

We have now completed our discussion of the thermal motions of the atoms. However, to obtain a complete description of the liquid, we also require wave functions to describe motion on a macroscopic scale, that is, to describe the liquid when we say that it is flowing. We consider the behaviour of the liquid at absolute zero, when it is entirely superfluid. As a simple example of such motion, the entire fluid may move as a body with velocity $\boldsymbol{v}_s$. In this case, the centre of gravity coordinates may be separated out, and the wave function written

$$\psi = \left\{\exp \mathrm{i}(m/\hbar)\boldsymbol{v}_s \cdot \sum_i \boldsymbol{R}_i\right\}\phi, \qquad (11.11)$$

where, as before, ϕ corresponds to the ground state at rest in the laboratory system of coordinates.

We now represent more complex flow situations in the liquid, where the velocity $\boldsymbol{v}_s(\boldsymbol{R})$ varies from point to point, but only very gradually on an atomic scale. If we consider the atoms in a small region about a point P, the motion of their centre of gravity may be allocated a velocity $\boldsymbol{v}_P$. These atoms will make a contribution to the phase term of (11.11) equal to

$$(m/\hbar)\boldsymbol{v}_P \cdot \sum_i \boldsymbol{R}_i,$$

the summation being taken over all the atoms in the region. Other regions will give corresponding contributions, so the total phase factor will be $(m/\hbar)\sum \boldsymbol{v}(\boldsymbol{R}_i)\cdot\boldsymbol{R}_i$, the summation now being taken over all the atoms of the liquid. Thus the wave function of the whole liquid will be of the form

$$\psi=\left\{\exp \mathrm{i}\sum_i s(\boldsymbol{R}_i)\right\}\phi, \tag{11.12}$$

where $s(\boldsymbol{R}_i)$ is some function of position. The above argument suggests that $s(\boldsymbol{R})$ is equal to $(m/\hbar)\boldsymbol{v}(\boldsymbol{R})\cdot\boldsymbol{R}$, the local velocity then being equal to the phase factor in (11.12) divided by $(m/\hbar)\boldsymbol{R}$. However, the wavelength of a wave will in general vary from point to point, so the velocity should be found by taking the gradient of the phase, namely

$$\boldsymbol{v}_\mathrm{s}(\boldsymbol{R})=(\hbar/m)\nabla s(\boldsymbol{R}), \tag{11.13}$$

and since the curl of the gradient of any scalar is identically zero,

$$\operatorname{curl}\boldsymbol{v}_\mathrm{s}=0. \tag{11.14}$$

12
ROTATION AND VORTEX LINES

12.1 The rotation of simply connected systems

In setting up the two-fluid model in Chapter 5 we included the relationship

$$\operatorname{curl} \boldsymbol{v}_s = 0, \tag{11.14}$$

and this condition imposes severe restrictions on the possible modes of motion. Consider a mass of helium in a simple cylindrical vessel or bucket which is rotating about its axis with a uniform angular velocity ω. To simplify matters we begin with temperatures sufficiently low that $\rho_s \simeq \rho$, so that to a good approximation $\boldsymbol{v}_s \simeq \boldsymbol{v}$, the velocity of the helium itself. The condition curl $\boldsymbol{v}_s = 0$ then implies that the helium must everywhere be at rest.

To see how the above result comes about, we introduce the circulation κ, which is defined as

$$\kappa = \oint \boldsymbol{v} \cdot \mathrm{d}\boldsymbol{l} \tag{12.1}$$

taken over any closed circuit in a fluid. According to Stokes' law, the circulation round any *infinitesimal* circuit in a liquid may be written

$$\oint \boldsymbol{v} \cdot \mathrm{d}\boldsymbol{l} = \oint \operatorname{curl} \boldsymbol{v} \cdot \mathrm{d}\boldsymbol{S}. \tag{12.2}$$

If we now expand the infinitesimal circuit via any path through the liquid, the form of (12.2) is such that it remains valid for the larger circuits so created. Alternatively we can say that (12.2) is valid for any *reducible* circuit in the liquid, that is for a circuit which can be shrunk to a point.

All possible circuits in a *simply connected region* as in a bucket, are reducible. (The liquid in the annular space between two concentric cylinders is *multiply connected,* and a circuit embracing the inner cylinder is not reducible.) Hence for helium in a simple bucket, eqn (12.2) and the condition curl $\boldsymbol{v}_s = 0$ seem to imply that

$$\oint \boldsymbol{v}_s \cdot \mathrm{d}\boldsymbol{l} = 0. \tag{12.3}$$

This condition of zero circulation can only be satisfied if $\boldsymbol{v}_s$ vanishes everywhere. This conclusion is in accord with Andronikashvili's experiment with a pile of oscillating discs (Section 5.5) where it appears that the superfluid does not take part in the rotational motion, but remains at rest. However, we now describe experiments which show that the whole mass of helium II in a bucket can be set into a state of uniform rotation.

In an experiment due to Osborne [139] a vertical glass tube was mounted between point bearings at top and bottom, and carried a small horizontal bar magnet. The assembly was rotated at speeds between 8 and 16 rev s^{-1} by an external magnetic field, and the profile of the meniscus viewed with a cathetometer through slits in the silvering of the dewar vessels. For a normal liquid the free surface would have a parabolic form given by

$$z = \frac{\omega^2}{2g} r^2,$$

where z and r are the vertical and radial coordinates of a point in the surface, and ω is the angular velocity. If only a fraction ρ_n/ρ of the helium rotates, then the centrifugal force will be reduced in this ratio, while the gravitational forces are of course unchanged. Hence at temperatures below 1.5 K the height of the meniscus should be reduced by a factor of at least ten. Osborne found that the helium II behaved as an ordinary liquid, and we conclude that both the normal and superfluids were rotating.

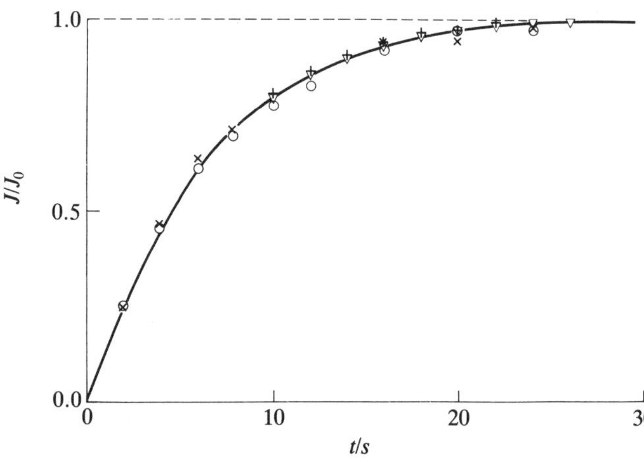

Fig. 12.1. The angular momentum J of a bucket of liquid helium, as a function of the time t after the start of rotation of the bucket at a speed of 0.78_5 rad s^{-1}. The value J_0 corresponds to complete rotation of the whole fluid (Hall [140].)

In Andronikashvili's experiment the superfluid between the pile of discs remained at rest during the oscillations. However, the position is different if the discs are steadily rotated for some seconds, for the whole mass of liquid is then set into rotation. Figure 12.1 shows the results of an experiment, due to Hall, in which a vessel of helium containing closely spaced discs was suddenly brought into rotation. The discs accelerated the helium until a steady state was reached. By measuring the torque on the system and integrating over the time, Hall obtained the total angular momentum imparted to the liquid. The experiment was performed at a temperature of about 1.27 K, where $\rho_n \simeq 0.04\rho$, and $\rho_s \simeq 0.96\rho$. Hence if both the normal and superfluids rotate, the imparted angular momentum will be about 24 times greater than if only the normal fluid rotates.

Figure 12.1 shows values derived for J, the angular momentum of the system, as a function of the time of rotation at an angular velocity of 0.78 rad s^{-1}. We see that after about 20 s the angular momentum rises to a value J_0 corresponding to complete rotation of the whole fluid.

12.2 The two-fluid model in rotating helium

The experiments described in the previous section show that the whole mass of liquid helium may be set into a state of uniform rotation. Hence the condition curl $\boldsymbol{v}_s = 0$, which we have hitherto taken as part of the two-fluid model, cannot be generally valid.

Rotation affects only some properties (for example, there is no change in either the fountain pressure [141] or the thermal excitation spectrum [142]). An informative experiment by Hall and Vinen [143] measured the velocity and attenuation of second sound in rotating helium using the resonators shown in Fig. 12.2. The resonator in Fig. 12.2(a) is circularly

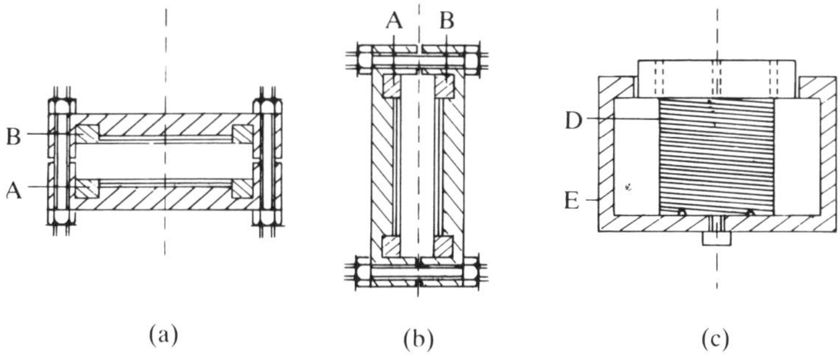

Fig. 12.2. Resonators used to observe the propagation of second sound in rotating helium. The axes of rotation are vertical. (a) and (b) show two orientations of an axial mode resonator and (c) a radial mode resonator. For further details see text (after Hall and Vinen [143].)

symmetrical about the vertical axis of rotation, and acts as a half-wave resonant cavity in which the second sound propagates parallel to the axis of symmetry. The heater generating the second sound is wound to and fro between slots in the ring A, and the phosphor bronze thermometer used as a detector is wound on ring B. The resonator in Fig. 12.2(b) is similar, but mounted so that the sound propagates perpendicularly to the axis of rotation. The third resonator (Fig. 12.2(c)) is a radial mode cavity, consisting of the annular space between an ebonite cylinder D and a brass outer case E. The resistance wires forming the heater and thermometer are wound on the cylinder D.

The amplitude of the signal in cavity 12.2(a) was apparently unchanged by steady rotations of the order of a few radians per second, but the signals in the other two cavities were appreciably reduced. The extra attenuation was a function of temperature, and *directly proportional to the speed of rotation.* In addition the coefficient of attenuation for a given velocity and temperature appeared to be the same for cavities (b) and (c).

Other experiments indicate that heat fluxes in helium II are considerably reduced by rotation, as was first shown by Kapitza. He observed the heat flow in helium II contained in the annular space between a glass rod of diameter 0.50 mm and a glass capillary tube of internal diameter 0.62 mm. With a given heat flux, the temperature difference across the ends of the tube (4 cm long) increased very considerably when the central rod was rotated.

12.3 Vortex lines

We now ask how the nature of liquid helium II is modified by rotation. Let us consider a cylindrical vessel of helium II, rotating at constant angular velocity ω, at so low a temperature that there is hardly any normal fluid, and we may ignore any distinction between the two densities ρ_s and ρ. For the same reason we will also ignore the distinction between the internal energy U and free energy F. We have seen that the helium has the normal angular momentum of an ordinary liquid, how then does the liquid helium behave so as to have the lowest energy characteristic of thermal equilibrium?

The usual condition for equilibrium in a system within a closed vessel rotating with a constant angular velocity $\boldsymbol{\omega}_0$ is obtained by minimizing the quantity $(U - \boldsymbol{I} \cdot \boldsymbol{\omega}_0)$, where U is the internal energy and $\boldsymbol{I}$ the total angular momentum of the system (see, for example Landau and Lifshitz [144]). For ordinary liquids this criterion leads to the expected result that the liquid rotates as a solid body with velocity $\boldsymbol{\omega}_0$. However, it is hard to see how the helium II can behave in this way, as it is virtually all superfluid and there are no viscous interactions between the atoms. We must therefore consider other ways in which the helium might rotate.

The experiments of Section 12.1 show that the whole mass of liquid helium II in a simply connected region may be set into a state of rotation. Hence the condition curl $\boldsymbol{v}_s = 0$, used in the two-fluid model to ensure the absence of viscosity in the superfluid, cannot be generally valid. Therefore several authors have developed the hypothesis that rotating helium is threaded by an array of vortex lines, as originally suggested by H. London and Onsager. Such vortex lines permit the liquid to have the angular momentum appropriate to solid body rotation, at the expense of a very small increase in the energy, while maintaining the condition curl $\boldsymbol{v}_s = 0$ over almost the whole volume.

A simple axisymmetric classical vortex line in the superfluid has a circulation

$$\kappa = \oint \boldsymbol{v}_s \cdot \mathrm{d}\boldsymbol{l} \tag{12.4}$$

where the integral is taken around any path enclosing the line, and the flow field is given by

$$|\boldsymbol{v}_s| = \frac{\kappa}{2\pi r}. \tag{12.5}$$

On paths which do not enclose a line,

$$\oint \boldsymbol{v}_s \cdot \mathrm{d}\boldsymbol{l} = 0$$

or, equivalently, curl $\boldsymbol{v}_s = 0$ as before. Each atom in such a flow has orbital angular momentum $m\kappa/2\pi$. The ideal vortex flow field has a singularity at $r = 0$, but in reality there will be a central core with a radius a_0 which can be roughly estimated by balancing the force due to the surface tension T against the Bernoulli force, *viz.*

$$\frac{2T}{a_0} = \tfrac{1}{2}\rho v^2 = \tfrac{1}{2}\rho\left(\frac{\kappa}{2\pi a_0}\right)^2.$$

There are reasons (see Section 12.4) for supposing that $\kappa = h/m$ and with this value we find

$$a_0 = \frac{\rho\kappa^2}{16\pi^2 T} \simeq 0.3 \text{ Å}.$$

Thus we expect no hole in the liquid, but rather a narrow region where the wave function tends to zero. The exact value of the energy parameter a_0 can only be found if we have a good wave function for the core, but it will clearly be of atomic dimensions.

12.4 The quantization of circulation

To obtain a complete description of a vortex line we need to specify its wave function. In the region of its core this is a very difficult problem which has not yet been solved. However, we can describe the irrotational flow pattern outside the core with Feynman's wave functions introduced in Section 11.7. In this case it follows that the circulation round a vortex line in helium II should be quantized (Feynman [137]).

Let us consider a system of helium at a sufficiently low temperature that we may ignore the excitations, so that its wave function is given by eqn (11.12),

$$\psi = \left\{\exp \mathrm{i}\sum_i s(\boldsymbol{R}_i)\right\}\phi, \tag{12.6}$$

where $\boldsymbol{R}_i$ specifies the position of the ith particle and $s(\boldsymbol{R}_i)$ is a function which changes steadily and smoothly with position. We now ask how this wave function changes if each atom in the ring of atoms shown in Fig. 12.3 moves a distance $\Delta\boldsymbol{R}_i$ into the position previously occupied by the $(i+1)$th atom. While making this rearrangement we temporarily displace such adjacent atoms as may be necessary in order to give a clear passage to the atoms in the ring. On account of the term ϕ, the wave function

Fig. 12.3. A displacement of each atom to the site of its neighbour in a ring is equivalent to a permutation of the atoms. For ^{4}He atoms the wave function cannot change as the result of such a displacement (Feynman [137].)

never falls to zero, so there are then no hidden changes of the sign in ψ during the displacement. Hence as all the atoms move one place round the ring, the phase must change by an amount

$$\sum_{\text{ring}} \{\text{grad}\, s(\boldsymbol{R}_i)\} \cdot \Delta \boldsymbol{R}_i .$$

According to (11.13) grad $s(\boldsymbol{R}_i) = (m/\hbar)\boldsymbol{v}_s(\boldsymbol{R}_i)$, so this change may be written

$$\oint (m/\hbar)\boldsymbol{v}_s(\boldsymbol{R}) \cdot \mathrm{d}\boldsymbol{R} = (m/\hbar)\kappa,$$

where κ is the circulation round the ring. As we are dealing with an assembly of Bose particles, the final state of the helium, in which each atom has taken the place of its neighbour, is indistinguishable from the initial state. Therefore the phase change must be zero or an integral multiple, n, of 2π. Hence

$$\kappa = \frac{nh}{m}$$

and the circulation is quantized. Moreover, h/m has a value of about $0.1\,\mu\text{J s kg}^{-1}$, so a single quantum is large enough to give rise to effects on a macroscopic scale.

The above theorem applies to any system of Bose particles, even at high temperatures where classical statistics apply. We appear to have proved too much, for there is no evidence that ordinary liquids exhibit a quantized circulation. Quantization of circulation is certainly implied by the wave function (11.12); the crucial point is whether this wave function adequately describes a particular liquid. Feynman's arguments given in Chapter 11 suggest that it is a good approximation for helium II.

We now describe, in this and the following section, two experiments which strongly suggest that the circulation in helium II is quantized. In the first, due to Vinen, the helium was contained within a metal tube, along the axis of which was stretched a fine metal wire (5 cm long and 0.025 mm diameter). For a perfectly uniform wire, in the absence of circulation, the modes of vibration of the wire are doubly degenerate, as it can vibrate independently in two perpendicular directions. However, any circulation κ round the wire removes the degeneracy, and the wire vibrates in two circularly polarized modes differing in frequency by an amount

$$\Delta f = \rho_s \kappa / 2\pi W,$$

where W is the sum of the mass per unit length of the wire and half the mass of liquid displaced by this length. The experiment consists in setting the helium into rotation and measuring the frequency splitting Δf.

Vinen found that some of his readings changed if the wire was vibrated very strongly, probably because a vortex line was only partly, and therefore unstably, connected to the wire. The values of all the stable readings are plotted in Fig. 12.4 as a histogram, and we see that most of the points correspond to a circulation of either zero or approximately h/m. This result has been confirmed in similar experiments by Karn *et al.* [146] who have observed circulations of up to $4\,h/m$

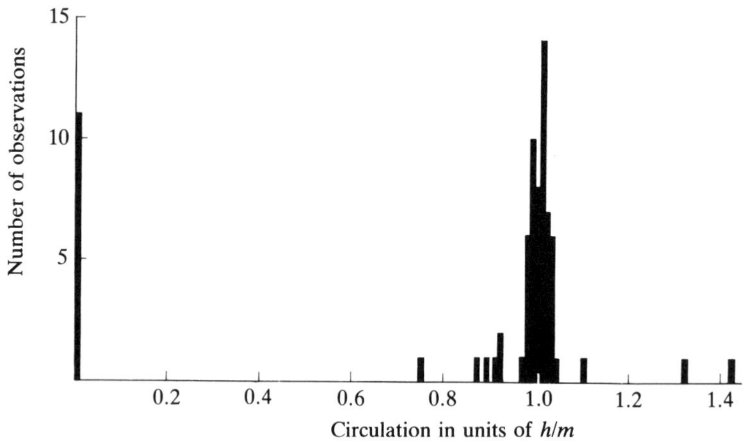

Fig. 12.4. Histogram showing the values of the circulation in Vinen's experiment which were stable against repeated vibration of the wire at large amplitude (Vinen [145]). More recently, quantized circulations have been observed with values up to $4\,h/m$ [146].

A notable feature of Vinen's experiment is that he observed the circulation to persist round the wire for several hours, and apparently indefinitely, after the cylindrical vessel containing the helium had been brought to rest. That is, a *persistent current* of circulating liquid can be set up in helium II analogous to a persistent current in a superconductor (see also references [36, 147]). Finally, we note that, as the circulation is quantized, a vortex motion must have a certain minimum energy. Therefore we might expect that at sufficiently low speeds of rotation, no quanta will be excited, and the superfluid will remain at rest. This effect has been demonstrated by Hess and Fairbank [148], but the critical speed is so extremely low that we may generally assume that rotation always produces vortices.

12.5 Ions and vortices

In Section 10.6 we described studies of the motion of ions in liquid helium II under the action of an electric field. In general the ions are scattered by the thermal excitations, so it is necessary to apply an electric field in order to maintain a uniform drift velocity. We now describe an experiment by Rayfield and Reif [149], using rather higher fields, at temperatures down to 0.3 K, where there are hardly any excitations to cause scattering. The ions then move very much as free particles with no retarding resistance.

Figure 12.5 shows the velocities of these ions as a function of the accelerating voltage V. The results show two remarkable features. The velocities are very much lower (order of 10 cm s^{-1}) than those observed in the experiments described above in Section 10.6. Moreover, the velocity of the ions *decreases* with increasing energy, a result at first sight so improbable that the authors describe confirmatory experiments with different electrode arrangements. They conclude from the very low velocities that the carriers of the electric charge are quite different entities from those in the high-temperature experiments; they must be much more massive. Moreover they travel more slowly as their energy

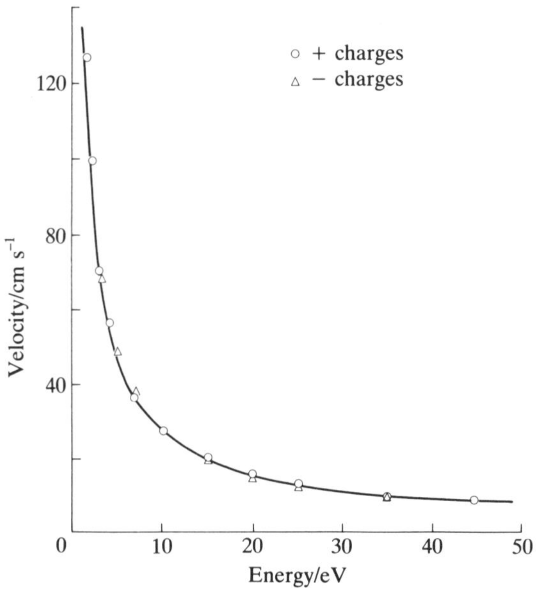

Fig. 12.5. The velocity of positive and negative ions in liquid helium measured as a function of their energy. The full line shows the theoretical values calculated from equations (12.7) and (12.8), by assuming that the ion carries one quantum of circulation and has a core radius of 1.2 Å (Rayfield and Reif [149].)

increases! Rayfield and Reif point out that this unusual behaviour may be accounted for if the charge carriers are vortex rings.

According to classical hydrodynamics, the energy and velocity of a vortex ring of radius r and circulation κ in a fluid of density ρ are given by

$$E = \tfrac{1}{2}\rho\kappa^2 r(\eta - \tfrac{7}{4}) \tag{12.7}$$

$$v = \kappa/4\pi r(\eta - \tfrac{1}{4}), \tag{12.8}$$

where

$$\eta = \ln(8r/a_0)$$

and a_0 is the radius of the core; note that η depends only weakly on r. It then follows from (12.7) and (12.8) that the velocity v of a vortex ring with circulation κ and energy E varies approximately as $1/E$, in accord with the experimental results. The exact relation between E and v depends on the two parameters κ and a_0. The authors therefore proceed by taking κ to have its quantized value h/m, and then choose a_0 to give the best fit to the experimental points. The best value comes out to be $a_0 = 1.2$ Å, and this leads to the theoretical curve shown in Fig. 12.5, which is in excellent agreement with the experimental points. Thus the experiment gives support to (a) the existence of vortex rings in helium II, (b) the fact that their circulation is close to the quantum value of h/m, and (c) a value of the parameter of a_0 close to 1.2 Å. For a discussion of more recent experiments to study the vortices by the ion technique see Schwarz [133] and Zoll [150].

12.6 Arrays of vortices

The most stable arrangement of vortices in a vessel of uniformly rotating helium will be a system of lines, parallel to the axis of rotation which minimizes the quantity $U - I\omega_0$. A detailed tabulation of the geometric form of these arrays has been given by Campbell and Ziff [151]. A simple estimate of the spacing of vortex lines for a given ω_0 may be obtained by making use of the known approximate uniformity of the arrays. In order to simulate solid-body rotation with angular speed ω_0, the circulation round any circular path of radius r concentric with the axis of rotation must be $2\pi r^2\omega_0$. However this total circulation must also be equal to $(\pi r^2)n_0 h/m$ where n_0 is the number of lines per unit area. Equating these expressions,

$$n_0 = 2\omega_0 m/h. \tag{12.9}$$

For an angular speed of 0.5 rad s^{-1}, n_0 is approximately 1000 cm^{-2}, corresponding to a spacing between lines of about 0.3 mm, in reasonable conformity with an experiment described below.

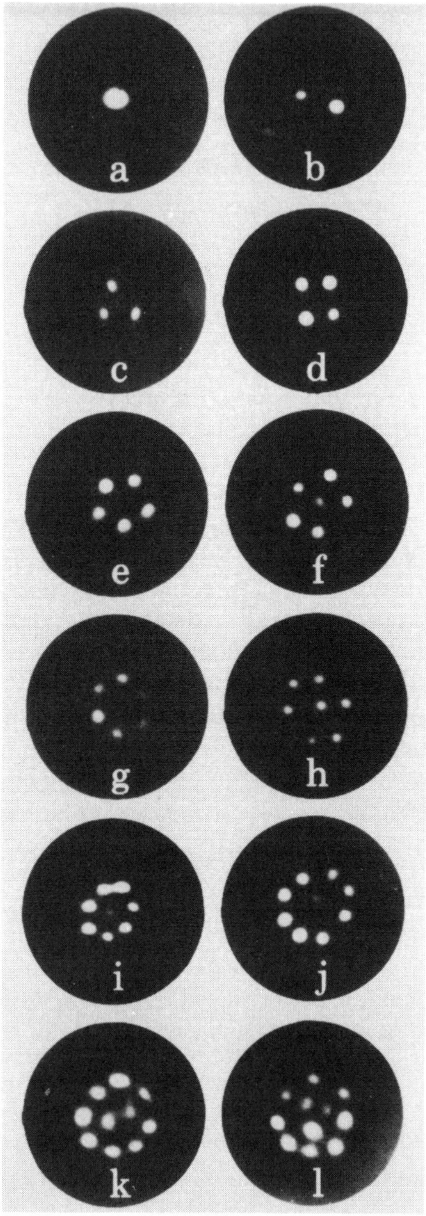

Fig. 12.6. 'Photographs' of quantized vortex lines in rotating ^{4}HeII, taken using a method described in the text. The circular aperture in each photograph is the cross section of the cylindrical container, of diameter 2 mm, and the angular speeds ranged from about 0.3 rad s^{-1} for one vortex line to a value two to three times larger for twelve vortex lines. From Yarmchuk *et al.* [152].

The presence and stability of vortex arrays in rotating ^{4}HeII, and the fact that the array rotates with the bucket, has been demonstrated in a fine experiment by Yarmchuk *et al.* [152]. These authors recorded the positions of the vortex lines by making use of the fact that negative ions (see Section 10.6) are trapped on vortex cores. The sample of ^{4}HeII filled a cylindrical bucket of 2 mm diameter, to a depth 25 mm. The cylinder's axis was placed at the rotation axis of a rotating dilution refrigerator (see Section 10.7) which maintained the sample at a temperature sufficiently low (100 mK) that electrons could pass freely through the vapour without scattering. At these temperatures there is virtually no normal component in the ^{4}HeII, so 0.8 per cent ^{3}He was added to provide some damping to maintain stability. The electrons trapped on the vortex cores were extracted, by application of an electric field, through the liquid meniscus (where the lines meet the free surface) and imaged on to a phosphor screen. The resulting light from the phosphor was conveyed through coherent fibre optics to room temperature, amplified in a low-light-level television camera, and recorded on a single frame of cinefilm. Some results are shown in Fig. 12.6.

12.7 Vortices in superfluid ^{3}He

Most properties of superfluid ^{3}He have been reviewed in Chapter 9, but the effects of rotation are better understood in the light of the behaviour of ^{4}HeII. The study of vorticity in ^{3}HeA and ^{3}HeB involves the rotation of buckets at temperatures below 3 mK and obviously presents formidable difficulties. However it now seems that vortices exist in both phases and that the vortex structures have much greater variety than in ^{4}HeII. In both phases the quantum of circulation is $h/2m_3 = 0.662 \times 10^{-3}\,\text{cm}^2\,\text{s}^{-1}$ where the factor 2 reflects the pairing of ^{3}He atoms. The cores are much larger than in ^{4}HeII because the coherence length ξ (roughly the size of a pair, or more strictly the range of a pair wave function) is of order 100 Å and spans many atomic spacings. The reader may pursue these matters further in references given by Hakonen *et al.* [153] and Volovik [154].

Salomaa and Volovik [155] have shown that in ^{3}HeB there may exist five types of axisymmetric vortices, differing in their internal symmetries. Two of these have normal (i.e. not superfluid) cores while the other three have superfluid cores. One of the superfluid cores is also ferromagnetic with a diameter of several thousands of Ångströms, and NMR effects attributed to this have been observed by Hakonen *et al.* [156]. There is also evidence that these cores are responsible for a first-order phase transition in rotating ^{3}HeB, whose position in the phase diagram has been detected by both NMR and gyroscope experiments [157].

In ^{3}HeA there is evidence for what is called continuous vorticity, an expression which is intended to convey the idea that the circulation,

$$\oint \boldsymbol{v} \cdot \mathrm{d}\boldsymbol{l},$$

is not restricted and may take *any* value, depending on the path chosen, with quantized values resulting *only* when particular paths are followed corresponding to primitive cell boundaries in the two-dimensional array of vortices. This feature, which is in marked contrast to the ^{4}HeII case, implies that ^{3}HeA does not obey curl $\boldsymbol{v}_s = 0$ and this is related to the existence of some textural dissipative effects which are probably also responsible for the difficulty of observing persistent currents in ^{3}HeA (see Chapter 9). The lattice of vortices exists as an equilibrium configuration only when there is rotation, and the area of each primitive cell depends on the angular speed ω as $Nh/4m_3\omega$ where N is the number of quanta $h/2m_3$ of circulation for a boundary path taken round one primitive cell (note that with $\omega = 0$ the cell size is theoretically infinite and the equilibrium configuration shows no vorticity). For a given ω, several different possible textures of the $\boldsymbol{l}$ and $\boldsymbol{d}$ vector fields (see Section 9.9) may be expected each with a different value of N. For example, the texture with one quantum of circulation ($N = 1$) has a singularity in the $\boldsymbol{l}$-field at each cell centre whereas that with two ($N = 2$) shows no singularity but instead a continuous field pattern.

13
VORTEX LINES IN HELIUM II

The experiments of the last chapter imply that a vessel of uniformly rotating helium II is threaded by an array of vortex lines. We now outline some properties of helium associated with the presence of these lines, and then discuss other experiments in which high velocities of the normal and superfluids produce a state of *turbulence*, in which the liquid is threaded by an irregular tangled mass of vortex lines. We begin by considering the appropriate density to be associated with the vortex motion.

13.1 Vortex lines and excitations

So far, we have only considered the rotation of helium II at temperatures close to 1 K where there is little distinction between the density of the superfluid and that of the actual liquid. We must now ask how the presence of thermal excitations affects the vortex lines.

It is often stated that at the higher temperatures the vortex lines are disturbances in the superfluid. Although this may be a convenient shorthand for expressing the true position, it is also a misleading one. The vortex motions at all temperatures are essentially motions of helium atoms in the real liquid. However, except at temperature zero, we have to consider the effect of the thermal excitations on this vortex motion. Following the method of Section 7.4, we fix our attention on a small element of volume in some liquid helium at temperature zero. Due to the vortex motions this element will have a certain velocity, say $\boldsymbol{v}_s$. We next raise the temperature so as to create a number of thermal excitations on which we impose a drift velocity $\boldsymbol{v}_n$. Then, as shown in Section 7.4, the momentum of the volume element may be written as $\rho_n \boldsymbol{v}_n + \rho_s \boldsymbol{v}_s$, where ρ_n and ρ_s are the densities of the normal and superfluids. That is, the density associated with the velocity field $\boldsymbol{v}_s$ is no longer ρ but ρ_s. To this extent the vortex is 'in the superfluid', but it is hard to form a physical picture.

The need to associate the superfluid density ρ_s with the vortex motion, rather than the total density ρ, is confirmed by experiments on the angular momentum of a vessel of rotating helium [158], and particularly of a gyroscope filled with the liquid [36, 159]. In one of these experiments the angular momentum of the helium in the gyroscope, itself immersed in a temperature bath of liquid helium, was sensed by applying a small torque to the system magnetically, and observing the tilt of the gyroscope. The helium in the gyroscope was set into rotation above the lambda point, then cooled to some temperature below the lambda point, and the angular momentum deduced from the observed tilts of up to 10°. Then, without any further rotation of the system, the temperature of the cryostat was raised, and measurements made at other temperatures. The results given in Fig. 13.1 show that the angular momentum varied directly as the superfluid density as the temperature in one particular experiment was raised from 1.24 K towards the lambda point, and then reduced again. Note that although the angular velocity of the gyroscope was changed solely by altering the temperature, conservation of angular momentum is not violated, because angular momentum could be exchanged with the helium in the temperature bath. Under the circumstances of these experiments, it is the circulation of the helium in the gyroscope which is conserved, not its angular momentum.

(Experiments similar to the above have also been performed with superfluid ^{3}HeA and ^{3}HeB [82, 83, 85] as described in Section 9.4. Currents have been observed to circulate undiminished for the order of a day, implying viscosities up to 12 orders of magnitude less than in the

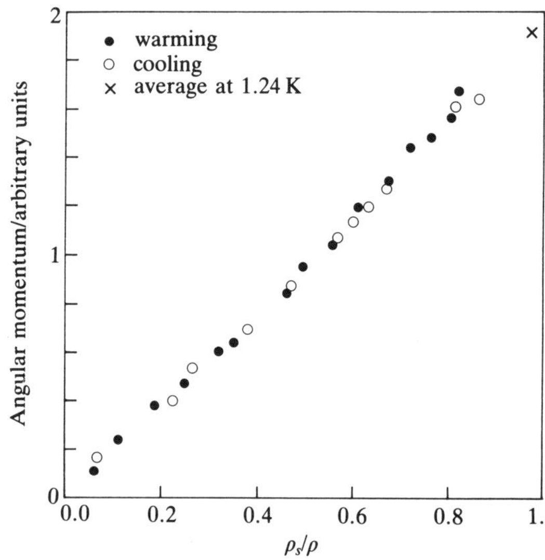

Fig. 13.1. Each point shows the angular momentum of persistent currents in liquid helium at a particular temperature, plotted against the values of ρ_s/ρ appropriate to that temperature (Mehl and Zimmermann [36].)

normal phase just above the transition temperature. For most purposes this means that $\eta_s = 0$. These results appear to be quite similar to those of Mehl and Zimmermann [36] for ^{4}HeII, in which η_n appeared to decrease by at least 11 orders of magnitude below the lambda temperature.)

13.2 Mutual friction

The core of a vortex is a very non-uniform region, and will therefore scatter the thermal excitations, i.e. the normal fluid. Hence, any motion of the normal fluid relative to a vortex line results in a force between the line and the fluid. The sums of these forces per unit volume give the so-called mutual friction F_{ns} between the normal and superfluids. It is this mutual friction term which is responsible for the attentuation of second sound in rotating helium described in Section 12.2.

Vinen has given a detailed discussion of the mutual friction [160, 161] but we shall here only outline some essential features that emerge from his analysis. An excitation moving parallel to a vortex line moves in a constant scattering potential, so the only scattering arises from excitations with a velocity component normal to the vortex line. This gives rise to a force per unit length of line

$$f = A\rho_n\rho_s(v_n - v_s),$$

where A is a constant which may be estimated, and the term $(v_n - v_s)$ takes account of the relative motion of the normal fluid and the vortex line. The total length of line in uniformly rotating helium is proportional to the angular velocity ω_0, hence the total mutual friction acting on unit volume of normal fluid

$$F_{ns} = A'\rho_n\rho_s(v_n - v_s)\omega_0, \tag{13.1}$$

where A' is another constant.

Given eqn (13.1) it follows from the equations of motion that there is an additional attenuation of second sound of the form $\alpha' = A'\rho\omega_0/2c_2$, where c_2 is the velocity of the sound. That is, there is an additional attenuation of the sound proportional to the angular velocity of rotation, as is observed. Moreover we expect the attenuation to be very small when the second sound is propagating parallel to the axis (i.e. when $\mathbf{v}_n - \mathbf{v}_s$ is parallel to the vortex lines), and to be the same for all directions of propagation perpendicular to the axis. This is, in fact, the symmetry observed in the experiments. It is difficult to calculate the constant A' precisely, but its value turns out to have the correct order of magnitude.

13.3 Vortex waves

The mutual friction force besides describing the attenuation of second sound in rotating helium, also gives an explanation of two experiments involving oscillating discs. The first is a variant of Andronikashvili's experiment due to Hall [162], in which a pile of rough discs was allowed to oscillate in a vessel of uniformly rotating ^{4}HeII at a low temperature (1.27 K) such that $\rho_s \approx \rho$. Rotation of the helium produced changes in the resonant frequency, and these changes were first discussed in terms of an increase $\Delta\rho$ in the density of the fluid dragged round by the discs. However, for some frequencies, $\Delta\rho$ was *negative*, a result quite inexplicable in terms of the normal fluid dragging some of the superfluid into motion. Also the value of $\Delta\rho$ appeared to vary periodically with the spacing between the discs, another result we would not expect to follow from an ordinary frictional force acting between the two fluids.

In the second experiment a single disc, roughened on its upper surface, was suspended on a torsion fibre, and oscillated a short distance below the surface of a bath of helium [Hall, 162]. As time went on, the level of the liquid in the beaker slowly rose, because of film flow from the outer vessel, and the period of oscillation varied as shown in Fig. 13.2.

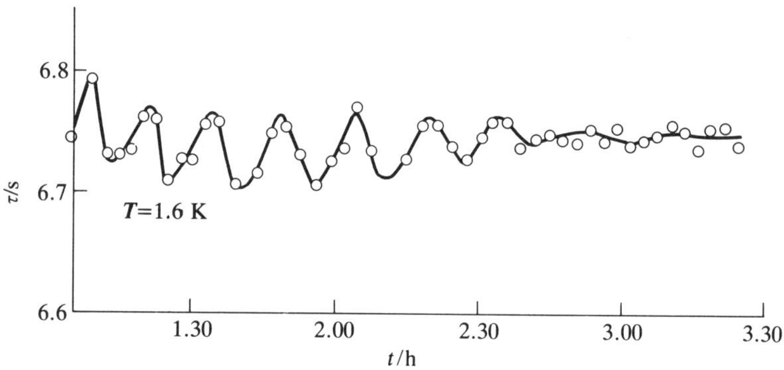

Fig. 13.2. The period τ of a disc oscillating in rotating liquid helium as a function of time (or of the changing levels of the liquid). $\omega_0 = 0.140\ \mathrm{rad\ s^{-1}}$ (Hall [162].)

These experiments are readily interpreted in terms of waves on vortex lines, as originally discussed by Lord Rayleigh in connection with ordinary liquids. As a vortex line has an energy per unit length, it will tend to reduce its length to lower its energy, that is, it appears to have a line tension. Hence if the vortex line is displaced in the liquid, so as to bow it out, the line tension will tend to return it to its equilibrium position. We thus have the necessary conditions for the line to support oscillations or vortex waves. In fact, the variations of frequency shown in Fig. 13.2 occur as the lengths of vortex line attached to the rough upper surface of the disc pass through resonant lengths for vortex waves.

13.4 Heat flow and turbulence

We described in Chapter 4 how the simple two-fluid model is only valid for not too large velocities of the two fluids. For example, the damping on a disc oscillating in ^{4}HeII is only independent of the angular amplitude until a certain value, when it increases rapidly. Similarly the heat flow in helium II is proportional to the temperature gradient for small gradients, but then falls off (Fig. 4.5). It seems that the two fluids experience additional resistance when moving at relatively high velocities. The onset of this resistance occurs at a *critical velocity,* orders of magnitude less than those given by Landau for the creation of thermal excitations. The rest of the chapter deals with the behaviour of the liquid in this *supracritical* region. The transition to the supracritical region is usually quite abrupt, as shown by the measurements of Fig. 4.6. We see that the extra resistance soon exceeds the initial value in the linear region. The heat flow then varies approximately as the cube root of the temperature gradient, as shown in Fig. 13.3. This additional resistance is not observed immediately on passing a heat flux through the liquid, but takes a few seconds to build up. Vinen studied this transition by passing a heat current transversely through waves of second sound generated in a resonator, and observing the amplitude of the second sound as a function of the heat current. He found that small heat flows made no difference to

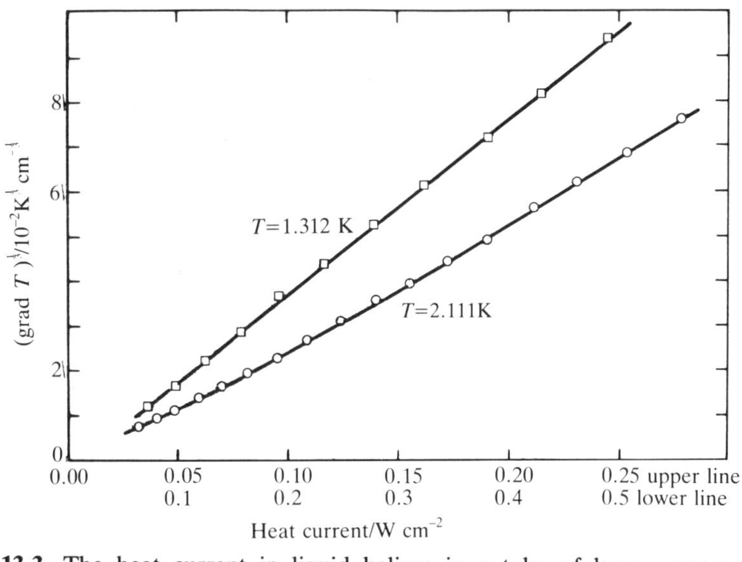

Fig. 13.3. The heat current in liquid helium in a tube of large cross-section (2.40 × 6.4 mm) plotted against the cube root of the temperature gradient (Vinen [160].)

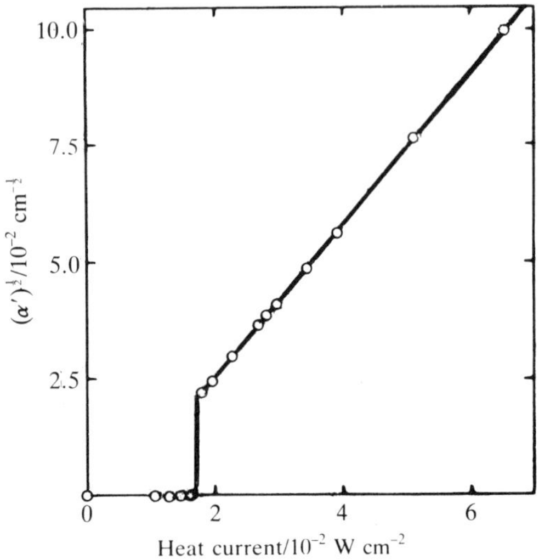

Fig. 13.4. The additional attenuation α' of second sound due to a heat current (Vinen [160].)

the amplitude of the second sound, but on exceeding a critical value, the attenuation of the second sound built up to an additional value roughly proportional to the square of the heat flux (Fig. 13.4).

The build-up times observed by Vinen were always reproducible provided that the liquid had been rested for several minutes prior to the measurement. However, observations of the second sound made a few seconds after switching off the heat current showed that the liquid was still disturbed. Hence, on switching on the heat current again, the thermal resistance and attenuation built up more quickly. Other authors have studied the temperature gradients in long capillaries filled with helium II. When the critical velocity is exceeded, the transition to the supracritical region is observed as a region of higher thermal resistance spreading out from one or more points on the tube. All these results are reminiscent of the behaviour of turbulence in ordinary liquids, and suggest that some form of turbulence is being created in the helium.

The turbulence impedes the flow of the normal fluid and hence gives rise to an additional thermal resistance. A qualitative analysis of this effect has been given by Vinen who assumes reasonably enough that the turbulent state is both homogeneous and isotropic. In addition, the growth of turbulence at any instant is assumed to depend only on the present state of the helium and not on its previous condition.

The length of vortex line per unit volume may increase either by 'stretching' or by 'production' processes occurring throughout the liquid.

By modelling the production process on the growth of a vortex ring Vinen [161] deduces that the rate of production of vortex line may be written in the form

$$\left[\frac{\mathrm{d}L}{\mathrm{d}t}\right]_{\mathrm{p}} = \frac{\chi_1 A''}{\kappa} \rho_{\mathrm{n}} L^{\frac{3}{2}} |\boldsymbol{v}_{\mathrm{n}} - \boldsymbol{v}_{\mathrm{s}}|$$

where χ_1 is a constant and A'' is related to the constants A and A' introduced in Section 13.2. Line is also being continuously destroyed, and Vinen models this process on the mutual annihilation of two parallel but oppositely directed vortex lines, and finds

$$\left[\frac{\mathrm{d}L}{\mathrm{d}t}\right]_{\mathrm{d}} = \frac{\chi_2 \kappa L^2}{2\pi},$$

where χ_2 is a constant. Equating the rates of production and destruction leads to an equilibrium value

$$L_0 = \gamma^2 \, |\boldsymbol{v}_{\mathrm{n}} - \boldsymbol{v}_{\mathrm{s}}|^2$$

where $\gamma = 2\chi_1 A'' \rho_{\mathrm{n}} \pi / \kappa^2 \chi_2$ is a function of temperature. The total force maintaining the equilibrium length L_0 of vortex line is proportional to L_0 and to $|\boldsymbol{v}_{\mathrm{n}} - \boldsymbol{v}_{\mathrm{s}}|$, and hence the retarding force on the normal fluid is

$$F_{\mathrm{ns}} \propto (v_{\mathrm{n}} - v_{\mathrm{s}})^3.$$

The equations of motion in Section 5.2 may now be extended by the inclusion of $\boldsymbol{F}_{\mathrm{ns}}$ and of a normal-component retarding force $\boldsymbol{F}_{\mathrm{n}}$ which in the linear region takes the form η_{n} curl curl $\boldsymbol{v}_{\mathrm{n}}$. For completeness we also include a superfluid-component retarding force $\boldsymbol{F}_{\mathrm{s}}$, although there has been some debate about the need for this. We thus obtain the steady state conditions

$$(\rho_{\mathrm{s}}/\rho)\,\mathrm{grad}\,p = \rho_{\mathrm{s}} S\,\mathrm{grad}\,T - \boldsymbol{F}_{\mathrm{ns}} - \boldsymbol{F}_{\mathrm{s}} \qquad \text{(superfluid)}$$

$$(\rho_{\mathrm{n}}/\rho)\,\mathrm{grad}\,p = -\rho_{\mathrm{s}} S\,\mathrm{grad}\,T + \boldsymbol{F}_{\mathrm{ns}} - \boldsymbol{F}_{\mathrm{n}} \qquad \text{(normal)}$$

which can be rearranged to give

$$\mathrm{grad}\,p = -\boldsymbol{F}_{\mathrm{n}} - \boldsymbol{F}_{\mathrm{s}} \tag{13.2}$$

and

$$\rho_{\mathrm{s}} S\,\mathrm{grad}\,T = (\rho_{\mathrm{n}}/\rho)\boldsymbol{F}_{\mathrm{s}} - (\rho_{\mathrm{s}}/\rho)\boldsymbol{F}_{\mathrm{n}} + \boldsymbol{F}_{\mathrm{ns}}. \tag{13.3}$$

In the linear region, $\boldsymbol{F}_{\mathrm{s}}$ and $\boldsymbol{F}_{\mathrm{ns}}$ are negligible, and in this case the heat flow is proportional to grad T as in Section 5.7. At higher flow rates, the motion is determined predominately by the term $\boldsymbol{F}_{\mathrm{ns}}$, which varies as $(v_{\mathrm{n}} - v_{\mathrm{s}})^3$, and results in a heat flow proportional to (grad $T)^{\frac{1}{3}}$ in agreement with the results shown in Fig. 13.3.

13.5 Self-sustaining vortex tangles

Vinen's ideas have been underpinned recently by detailed computations by Schwarz [163, 164] based on computer modelling of interacting vortex lines. The lines become curved by interaction with the flow, turn into loops and multiply by interconnecting with themselves or with other lines, and are annihilated by interaction with the walls. In the computer experiment, a sample volume consisting of a length of the channel is filled with an arbitrary initial vortex configuration and subjected to a uniform driving velocity v_0 along the channel. The development of this configuration as it interacts both with the channel walls and itself is then calculated as a function of time. It is found that the vortex tangle in the channel reaches a well-defined steady state corresponding to computable values of L_0 and γ, which may be compared with values deduced from experiment [165, 166]. For example, Ashton *et al.* [165] immobilized the normal component by using a narrow tube, and thus measured v_s along a capillary by direct observation of the rate of change of the level difference between two baths. Then by also measuring grad T they deduced $\boldsymbol{F}_{ns}$ from eqn (13.3), $\boldsymbol{F}_n$ being zero in their experiments because

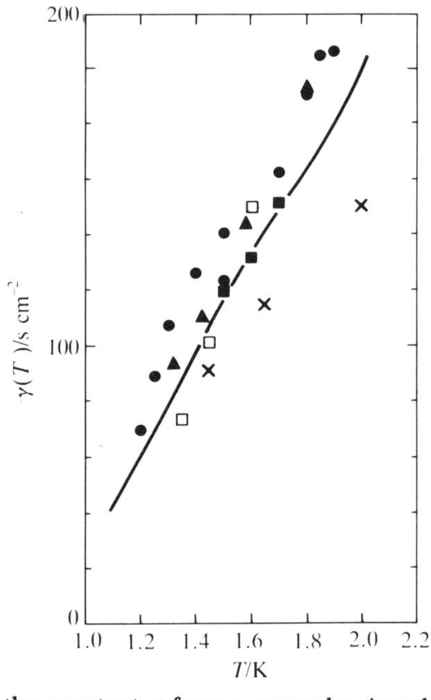

Fig. 13.5. Values of the constant γ from a paper by Awschalom *et al.* [166]. The experimental results are shown as points and the calculated values as a continuous line.

$\boldsymbol{v}_n = 0$, and $\boldsymbol{F}_s$ being taken as negligible. These data, together with information on A or A' (eqn (13.1)) from the attenuation of second sound in rotating helium, then allows γ to be derived. Figure 13.5 shows a selection of data from Ashton *et al.* [165], Awschalom *et al.* [166] and other authors for $\gamma(T)$.

14
CRITICAL VELOCITIES

In this chapter we consider the values of the critical velocities observed in different kinds of experiments not yet fully understood. (A good review of the situation up to about 1968 has been given by Keller [167]). As a background to the problems of critical velocities we recall the results of the previous chapter that the turbulent state may be described in terms of a mass of vortex line, whose total length increases with the increasing relative velocity of the two fluids. If there is *any* vortex line present to start with, then a sufficient flow rate can create more by the broadly understood dynamical rules governing the behaviour of vortices. We now ask where the first vortex comes from. Awschalom and Schwarz [168] have made sensitive ion-trapping experiments and conclude that any container of superfluid ^{4}HeII, handled by conventional methods, will be permeated from the beginning by numerous quantized vortices stabilized by pinning to solid surfaces. Such vortices are described as *remanent* vortices. Thus it seems probable that all experiments so far performed on the growth of quantized vortex lines have been influenced by some remanent vorticity.

14.1 Several critical velocities

Landau's criterion for flow without any generation of excitations, eqn (7.4), sets an upper limit of about 60 m s^{-1} at the vapour pressure, falling to about 46 m s^{-1} at higher pressures as shown in Fig. 10.12. These speeds have been observed in experiments using negative ions, the limit then being set by the emission of roton pairs (see Section 10.6). Other experiments [169] with a small orifice protected from vorticity by porous filters gave rise to critical velocities of order 10 m s^{-1}, limited apparently by the creation of vortex rings. Usually, however, the onset of turbulence destroys superfluidity at much lower speeds, often no higher than a few mm s^{-1}.

Some experiments show two critical velocities. Figure 14.1 shows the damping on a sphere oscillating in liquid helium II as a function of the maximum amplitude (or velocity) of the oscillation. At low amplitudes, region A, the damping has a constant value equal to that we would

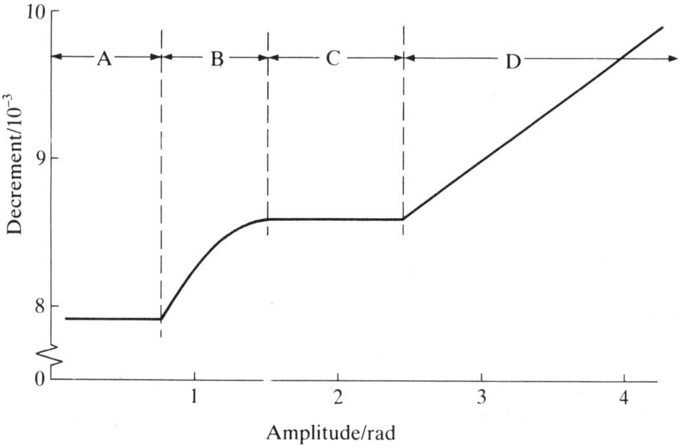

Fig. 14.1. The damping on a sphere oscillating in liquid helium at 2.149 K with a period of 18.5 s (after Donnelly and Hollis Hallett [170].)

associate with the normal viscosity η_n. At higher amplitudes there appear to be two critical velocities, at each of which the damping rises sharply. In the intermediate region *C*, the damping is independent of the amplitude, and is approximately what we would expect for an ordinary liquid of viscosity η_n and a density ρ equal to the total density of the helium. It appears that, above the transition at the lower amplitude, both fluids are coupled together by the mutual friction and move as one.

The first transition is probably due to the onset of turbulence in the superfluid component. The behaviour in this region is very complex as set out in an extensive review by Tough [171]. It seems that, depending on the various experimental conditions, there may be as many as four distinct turbulent flow régimes. We discuss some of the simpler aspects in the next section.

The second transition in Fig. 14.1 appears to correspond, at least approximately, with the onset of turbulence in an ordinary classical liquid with a viscosity η_n and density ρ, equal to the density of the whole liquid. That is, the Reynolds numbers for these transitions are quite similar to those observed in experiments on liquid helium I. This result again suggests that, in the region *C*, the normal and superfluids are coupled together quite closely. For the rest of the chapter, however, we shall be concerned only with the first critical velocity which marks the boundary of the pure superfluid condition.

14.2 Measurements of the critical velocity

The transition to the turbulent state exhibits itself in a wide variety of experiments. In general the critical velocity in flow experiments *increases*

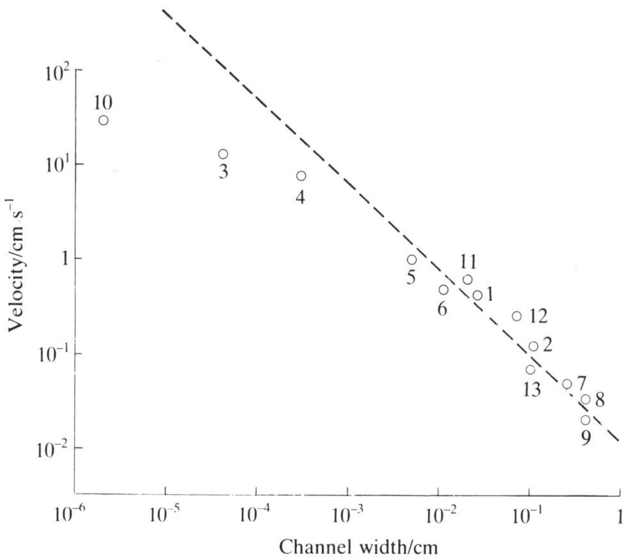

Fig. 14.2. The superfluid critical velocity v_c at a temperature of about 1.4 K as a function of channel size, measured in experiments on flow (1, 2), heat flow (3–6), second sound (7–9), film flow (10), oscillations (11, 12), and rotation viscometer (13). The dashed line shows the critical velocity given by eqn (14.2).

as the diameter of the channel *decreases*. In addition the observed critical velocities may also depend somewhat on the temperature. We now collect together the results of various experiments exhibiting a critical velocity v_c, and in view of the dependence of v_c on temperature we restrict ourselves to one temperature, 1.4 K, so chosen because most authors quote values hereabout.

Figure 14.2 shows the velocity of the superfluid at the onset of turbulence plotted against the diameter of the channel in mass and heat flow experiments, or against some other appropriate linear dimension. For example, the superfluid in a helium film flows with a velocity of the order of 30 cm s^{-1}; this presumably represents a critical velocity related to the thickness of the film ($\sim$20 nm). The relevant dimension in measurements on oscillating systems is probably the penetration depth for viscous waves. All these experiments give values for v_c in rough accord with each other.

Attempts to account for the observed critical velocities have in the past concentrated on producing some form of universal curve giving the velocity as a function of the width of the relevant channel. Two attempts described in the next section, based on criteria for the formation of vortices in the superfluid, lead to the dotted line shown in Fig. 14.2. The

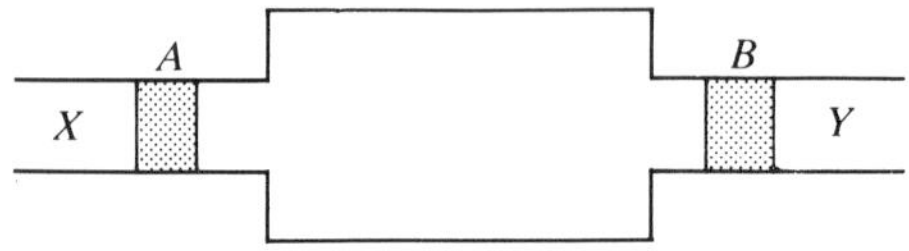

Fig. 14.3. Schematic diagram of apparatus to measure the critical velocity of superfluid flow when the normal fluid is clamped by constrictions at A and B; see text.

agreement with experiment is only fair, and there are fundamental difficulties to the theory which we outline in the next section.

More information on critical velocities has been given by experiments of a type shown very schematically in Fig. 14.3. The plugs at A and B consist of tightly packed powder, so that if a hydrostatic pressure is applied across the ends of the tube XY, the only flow possible in the central region will be motion of the superfluid, as the normal fluid is 'clamped' between the two plugs. De Bruyn Ouboter *et al.* [172] have described several ingenious experiments which have been used to measure the critical velocity of superfluid flow under such conditions. In particular, they have determined the critical velocity in the central portion of the tube of Fig. 14.3 by observing the rise of temperature associated with the onset of dissipative effects. The results of such measurements are displayed in Fig. 14.4, which gives the product of critical velocity and channel width as a function of channel width.

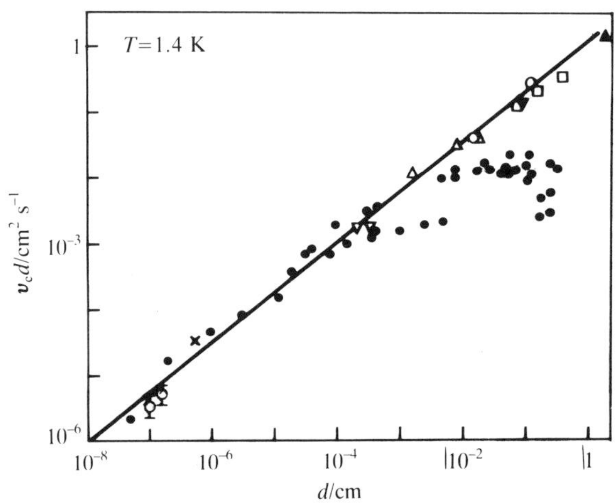

Fig. 14.4. The critical superfluid velocity v_c plotted as $v_c d$ vs. d, where d is the width of the relevant channel. The small solid circles are earlier results such as those shown in Fig. 14.2. The larger symbols were obtained with the normal fluid 'clamped' by plugs. The straight lines show the relation $v_c \simeq 1/\sqrt[4]{d}$ cm s^{-1}. (De Bruyn Ouboter *et al.* [172].)

Figure 14.4 shows that, for channel sizes above about $1\,\mu\text{m}$ in diameter, the critical velocities when the normal fluid is clamped are appreciably greater than those observed by other methods. This discrepancy rapidly becomes more marked with increasing channel size. In addition, the observed critical velocities are now independent of the temperature. All the velocities obtained in this way fall quite close to the straight line representing the relation $v_c \simeq 1/\sqrt[4]{d}\ \text{cm s}^{-1}$, where d is the width of the channel.

14.3 Theories of the critical velocity

Despite a mass of experimental results on turbulence in liquid ^{4}HeII (reviewed in [171]) there is still no completely satisfactory account of the conditions which determine the onset of turbulence. One of the latest accounts is that given by Schwarz [173] using his computer model (Section 13.5) to show that as the flow in a channel is reduced below a critical velocity the rate of production of vortex line becomes insufficient to balance the loss at the walls and the vortex tangle dissolves. This critical velocity has been investigated in a real experiment by Baehr *et al.* [174] using a flow channel with the normal component immobilized so that $\boldsymbol{v}_n = 0$. The technique was the same as that used by Ashton *et al.* [165] to obtain the results shown in Fig. 13.5, but lower values of $\boldsymbol{v}_s$ were used in a deliberate attempt to observe critical velocity effects. The results of this experiment using a channel of diameter 1.34×10^{-2} cm are shown in Fig. 14.5, and lie within the scatter of Figs. 14.2 and 14.4. The general magnitude of the critical velocities, about $15\ \text{mm s}^{-1}$, is in satisfactory agreement with theory, but the velocities are almost independent of temperature although the Schwarz model is based on the idea that the turbulent state is maintained by the *relative* velocity $(\boldsymbol{v}_n - \boldsymbol{v}_s)$ and this leads to the temperature dependence shown by the continuous line in Fig. 14.5(b). Until this problem is resolved the model must be used with caution.

Although current theory favours $(\boldsymbol{v}_n - \boldsymbol{v}_s)$ as the relevant driving term for the production of turbulence, there is at least one experiment [175] in which the onset of turbulence appeared to be governed by $\boldsymbol{v}_s$ rather than $(\boldsymbol{v}_n - \boldsymbol{v}_s)$. Hartoog and van Beelen [176] have observed a rather low value of $\boldsymbol{v}_s$ ($\sim 1\ \text{mm s}^{-1}$) at which there seems to be an onset of superfluid turbulence. Related to this is the question of whether the superfluid equation of motion should also include the term F_s of eqns (13.2) and (13.3). If so, then in principle it can be measured by using eqn (13.2) to interpret measurements of grad p in tubes where the normal component is immobilized so that $\boldsymbol{F}_n = 0$. However, these are difficult experiments

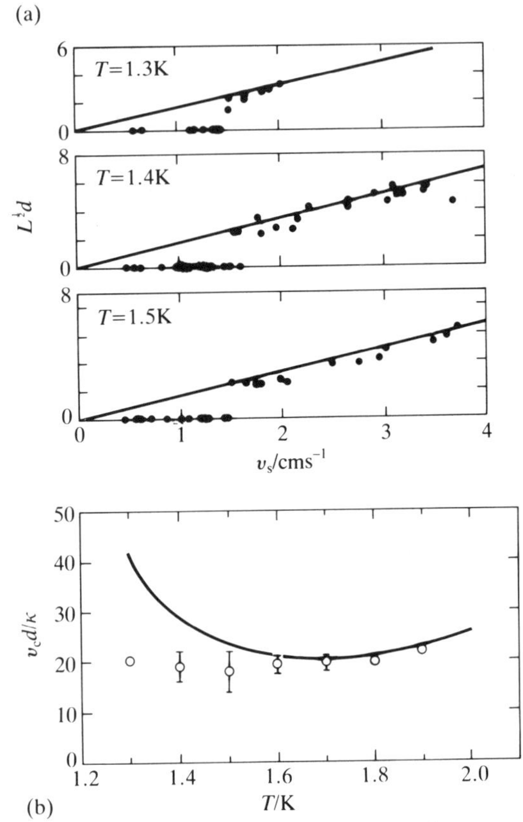

Fig. 14.5. (a) The vortex-line density L plotted as $L^{\frac{1}{2}}d$ against the superfluid velocity v_s showing the transition between friction-free and turbulent flow. (d is the inner diameter of the tube, 0.134 mm.) (b) The critical superfluid velocity v_c plotted as $v_c d/\kappa$ measured at various temperatures. The temperature dependence predicted by the Schwarz model is shown by the solid line. ($\kappa \equiv h/m_4 =$ 9.97 mm s^{-1}). From Baehr *et al.* [174].

because the pressure gradients are small and their interpretation is open to dispute [177].

There seems still to be no simple explanation of the empirical (diameter)$^{-\frac{1}{4}}$ dependence; it would be rather easier to understand a (diameter)$^{-1}$ dependence [178] arising from the known properties of vortex lines. De Bruyn Ouboter *et al.* [172] suggest that in experiments in which the normal component is *not* immobilized, the onset of classical turbulence in the normal component may show up in the form of lower critical velocities as for example with the data shown as small solid circles in Fig. 14.4.

Classical turbulence is associated with a critical velocity expressed in a dimensionless form as the Reynolds number. There have been many speculations about how the several critical conditions (see for example, Fig. 14.1) may be interpreted. There is also a range of suggested forms for the appropriate Reynolds number in terms of the various densities, viscosities, and velocities. Tough [171] lists seven such forms and shows them all to be unsatisfactory for the explanation of most data!

14.4 Vortex rings

There seems little doubt that vortex tangles are responsible for at least some of the complicated problems mentioned in this Chapter and that Schwarz's computational modelling is broadly the right way to proceed in those cases. But there are other situations [169, 178] where the behaviour of simple circular vortex rings may be the key to understanding. It is no longer possible to believe that vortex rings can give a complete explanation of critical velocities but the idea has an honourable history and we outline it here.

Atkins supposed that the vorticity in a capillary is initiated in the form of a closed ring of vortex line of radius R. He then regarded the ring as a particular form of elementary excitation, and applied the Landau criterion for the critical velocity

$$v_c = \epsilon / p,$$

where ϵ is the energy of the excitation and p its momentum. In fact a vortex ring carries no momentum, and the quantity must be replaced by I, the integrated impulse needed to create a vortex in a liquid initially at rest, see reference [179]. The energy and impulse of a ring of radius r are given by classical hydrodynamics as

$$E = \tfrac{1}{2}\rho\kappa^2 r\left\{\ln\left(\frac{8r}{a_0}\right) - \frac{7}{4}\right\}$$

$$I = \pi\rho\kappa r^2,$$

where κ is the circulation and a_0 the radius of the core. Substituting ρ_s for ρ, and the quantum value of h/m for κ, it follows that

$$\frac{E}{I} = \frac{\hbar}{mr}\left\{\ln\left(\frac{8r}{a_0}\right) - \frac{7}{4}\right\}.$$

As the logarithmic term depends only weakly on r, the lowest value of E/I will be given approximately by the largest permitted value of r, hence

$$v_c \approx \frac{\hbar}{mr_{\max}} \ln\left(\frac{1.4r_{\max}}{a_0}\right). \tag{14.1}$$

A quite similar expression has been derived by Feynman, who considers a flow of helium from the end of a tube into a vessel of still liquid.

Expression (14.1) has the merit that it gives some explanation for the magnitude of the critical velocity and its dependence on channel size. The value of a_0 is not critical and we use the figure of 1.2 Å deduced by Rayfield and Reif (Section 12.5). Then, assuming somewhat arbitrarily that the critical diameter of the ring $2r_{\max}$ is half the width of the channel d, we find that

$$v_c \approx \frac{4\hbar}{md} \ln\left(\frac{d}{3a_0}\right), \tag{14.2}$$

The relation (14.2) is shown by the dashed line in Fig. 14.2. We see that the expression gives a fair account of the critical velocity in relatively wide channels, but not when the width falls below 1 μm. A more fundamental difficulty is that expression (14.2) is derived by assuming that the vortex ring, when formed, is approximately as large as the diameter of the capillary. That is, it may be of the order of 1 mm in diameter. On the atomic scale, this is an enormous size, and it is hard to see how such a large co-operative motion can suddenly be initiated in the helium. A reasonable figure for the greatest diameter of a vortex ring which could be created instantaneously in the liquid is probably of the order of 10^{-6} cm, and this corresponds to a critical velocity of about 100 cm s^{-1}! It was this difficulty that led Vinen [180] to suggest the existence of remanent vortex lines which are caused to expand by the flow, setting up turbulence. Glaberson and Donnelly [181] went on to consider the behaviour of metastable lines pinned to surfaces and produced an expression for v_c:

$$v_c \approx \frac{\hbar}{md}\left[\ln\left(\frac{4d}{a_0}\right) - \frac{1}{4}\right]$$

which is quite similar to (14.2). For a recent review of the current position in this approach, and analysis of data on friction and drag on vortices, see Barenghi *et al.* [182].

15
MISCELLANEOUS ASPECTS OF LIQUID HELIUM II

We now briefly review some other features of liquid helium II of particular interest.

15.1 The microscopic theory of liquid ^{4}He

Feynman's derivation of the excitation spectrum of liquid ^{4}He, described in Chapter 11, postulates the general form of the wave function, and then uses a variational method to obtain a more detailed expression. Many attempts have since been made to derive the energy spectrum directly, by solving a Schrödinger equation. It is comparatively simple to write down such an equation for the liquid, but very difficult to obtain a rigorous solution. There is now a large literature on this problem, and we shall do no more than indicate one or two salient features.

The essential step in deriving the energy spectrum from the Schrödinger equation was made by Bogoliubov [183] many years ago, in a rather difficult paper which was virtually ignored for several years. Bogoliubov considered a dilute gas of Bose particles *with weak interactions*. Because the particles obey Bose–Einstein statistics, nearly all the atoms are condensed in the ground state at temperatures close to zero. Hence the number of particles in the ground state is macroscopically large, and this permits certain approximations to be made. Bogoliubov was then able to solve the Schrödinger equation, written in terms of creation and destruction operators, by means of a suitable transformation. It turns out that the excitations of low wave number no longer have the usual particle type spectrum $\epsilon = p^2/2m$, but the spectrum $\epsilon = pc$, where c is the velocity of sound. Thus the presence of the interactions changes the low energy particle-like excitations into phonons! This result has since been confirmed by many other authors.

The ideas involved in Bogoliubov's treatment of a dilute Bose gas with interactions are somewhat unusual. Thus, the treatment implies that if we start with a perfect Bose gas in the ground state, and introduce interactions, then some atoms will be scattered out of the zero-momentum state. If we now introduce one *excitation* of low momentum

p, there is a further increase in the number of atoms scattered out of the zero-momentum state. According to Brueckner and Sawada [183a], the average number of these extra particles at any instant has the form

$$(\hbar/p)(\alpha N m)^{\frac{1}{2}},$$

where N is the total number of atoms, m the atomic mass, and α a scattering parameter. Also, the fluctuation about this mean value is

$$(\sqrt{3}\hbar/2p)(\alpha N m)^{\frac{1}{2}}.$$

Hence, for a low value of p, both the average number of extra particles excited out of the zero-momentum state, and the fluctuations in this number, will be large. The excited state bears little resemblance to a single-particle state. It is, in fact, a phonon state.

The systems considered above are all much less dense than real liquid helium, and do not reproduce the characteristic minimum in the phonon-roton excitation curve. Therefore attempts have been made to extend these treatments to higher densities with stronger interactions. In this case the so-called depletion effect becomes important. With stronger

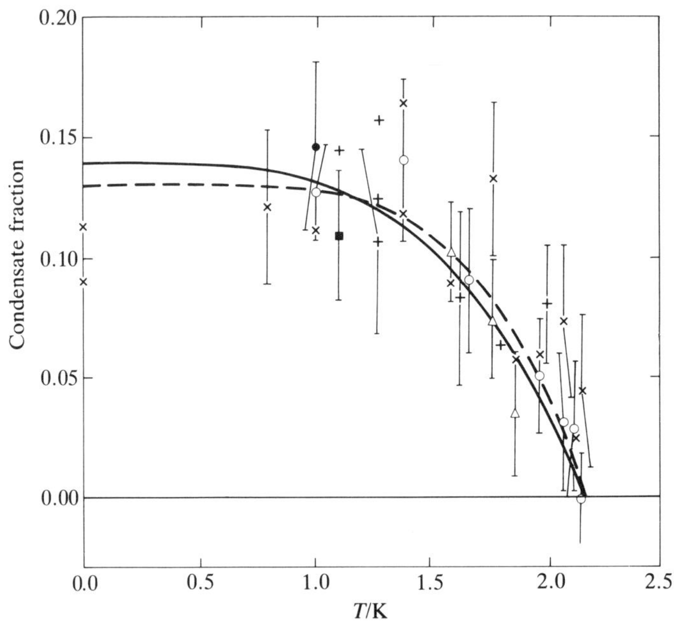

Fig. 15.1. Various esimates of the condensate fraction, deduced from neutron scattering experiments and from measurements of surface tension, compared with two theoretical curves. From Svensson [184a], who also gives all the detailed references.

interactions, more atoms are scattered out of the ground state, and Bogoliubov's approximation that virtually all the atoms are in the ground state is no longer valid. This depletion of the ground state may be quite large, and must be considered in any further calculations. The backflow round a moving atom (Section 11.5) must also be taken into account, but we do not pursue these points any further.

It is tempting to identify the condensed state, described above, with the superfluid component of the two-fluid model. Penrose [184] extended the work of Penrose and Onsager to non-zero temperatures at which phonon excitations are present, and identified the velocity $\boldsymbol{v}_s$ with the velocity of the state into which the condensation occurs. The present position has been reviewed by Svensson [184a]. Figure 15.1 shows the results of some recent experiments (of different kinds) and calculations which suggest that as T approaches zero, the fraction of atoms in the zero-momentum state approaches a value of about 13 per cent. In apparent contrast to these results, experiments to measure the superfluid density of Landau's theory of excitations in a superfluid background all agree that ρ_s/ρ approaches 100 per cent as T approaches zero (see Fig. 7.2). It is hard to reconcile these two values (100 and 13 per cent), and one cannot immediately identify the condensate with the superfluid component of the two-fluid model. More work is needed to reach a better understanding of this important point.

15.2 Josephson effects in helium

It was pointed out by Richards and Anderson [185, 186] that a description of superfluid helium in terms of a definite wave function with a definable phase Φ in every part of the system implied that Josephson effects, analogous to those found in superconductors [187] might be observable. This approach stresses Beliaev's result [188] that the wave function of the zero-momentum state in superfluid helium contains a phase factor $\exp(-i\mu t/\hbar)$, where μ is the chemical potential of an atom. They consider the experimental arrangement shown schematically in Fig. 15.2. Two baths of helium standing at different levels are connected by a small orifice. The chemical potentials of the helium at A and B will differ by an amount mgz where m is the mass of the atom and z the difference in hydrostatic level. Hence the phase of the helium at A will slip with respect to the phase at B, at a rate $\omega = mgz/\hbar$. Richards and Anderson suggest that this situation may be stabilized by the motion of vortices either across, or away from, the orifice. As the phase change round a vortex of n quanta of circulation is $2n\pi$, the condition for stability might be

$$mgz/\hbar = \nu \cdot 2n\pi$$

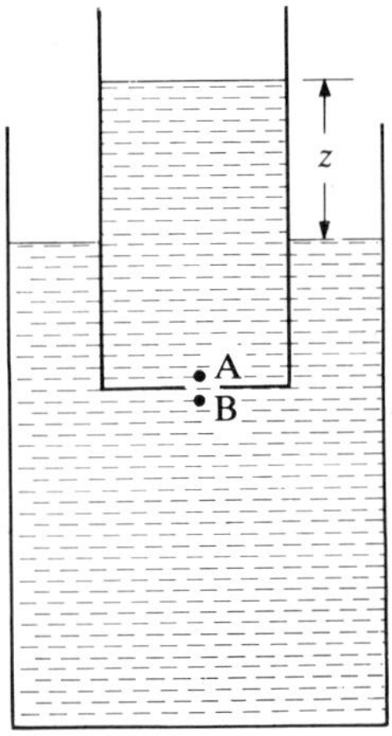

Fig. 15.2. Diagram to illustrate phase slip in liquid helium II (Richards and Anderson [185]). The two compartments are connected by a fine orifice (or weak link); see text.

or

$$z = nhv/mg, \tag{15.1}$$

where v is the rate at which vortices cross or leave the orifice.

Richards and Anderson sought to demonstrate this stability by introducing a transducer vibrating below the orifice at a frequency f, at an amplitude sufficient to induce turbulence in the liquid. The velocity field set up by the transducer lowers the level of the liquid in the central tube, both above and below the lambda point, in accord with the Pitot tube effect. Below the lambda point, however, the change in level appeared to occur in a series of steps. Taking v as being equal to the frequency of the transducer, it seemed that the steps occurred at intervals of (hv/mg). Although each step was not observed on every run, the experimental results certainly suggested a tendency towards such a spacing. (There is also a complication in that the vortices may be formed at a sub-multiple of the transducer frequency.) An experiment by Khorana and Chandrasekhar [189] seemed to confirm Anderson's prediction, with well-defined steps, about 1 mm high, in good numerical agreement with the value of hv/m_4g. Unfortunately the situation has become less clear with the strong suggestion that a number of investigators were in fact studying ordinary acoustic resonances rather than the A.C. Josephson effect, as discussed by Anderson and Richards [190]. In spite of this setback, work continues both in ^{4}HeII [191] and superfluid ^{3}He [192] but there are considerable experimental difficulties to overcome.

15.3 The specific heat near the lambda transition

The regime of liquid helium II with its unusual properties sets in very abruptly at the lambda point, 2.17 K. The first classic measurements very close to the lambda temperature were those of the specific heat by Buckingham *et al.* [193]. Their method was deceptively simple; they supplied heat to a specimen of helium at a constant rate and observed the change of temperature. In fact, by ensuring that thermal equilibrium was attained rapidly within the calorimeter, and that any heat leaks to the specimen were small and regular, they achieved a greater sensitivity than hitherto obtained in any specific heat measurements on any material. The thermal relaxation time for a mass of liquid helium I may be very large, because it has a high thermal capacity and a low thermal conductivity. The inside of the calorimeter therefore contained a system of copper fins, which ensured that none of the helium was further than 0.075 mm from a copper surface, and that the thermal relaxation time was less than 1s.

The results of Buckingham *et al.* are given in Fig. 15.3 on successively expanded temperature scales and show that the specific heat continues to rise as the temperature continues to approach the lambda point. The data was fitted empirically to an expression including a symmetrical logarithmic term $\log_{10} t$ where $t \equiv |T/T_\lambda - 1|$. More recent experiments [194] take account of two correction terms which arise as follows. At most

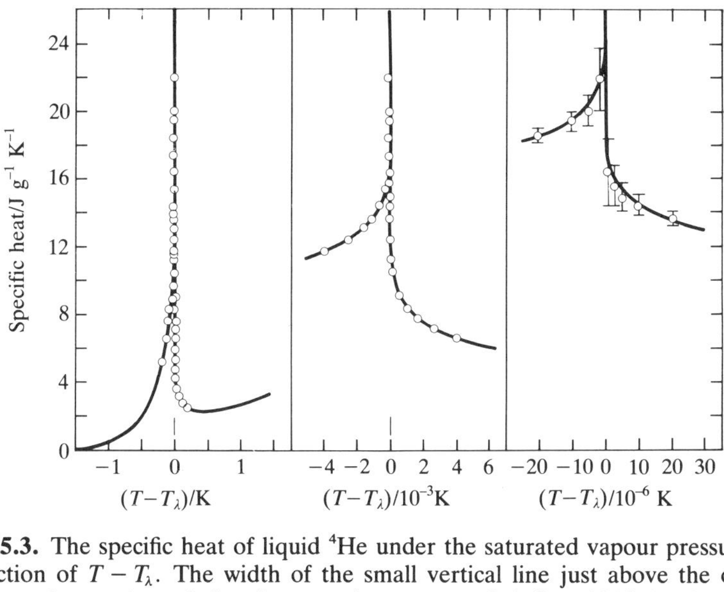

Fig. 15.3. The specific heat of liquid ^{4}He under the saturated vapour pressure as a function of $T - T_\lambda$. The width of the small vertical line just above the origin indicates the portion of the diagram shown expanded (in width) in the curve directly to the right (after Buckingham and Fairbank [193].)

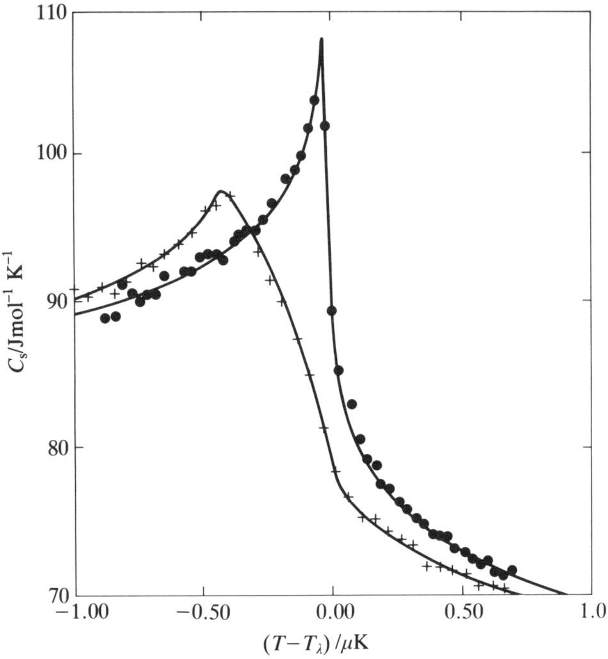

Fig. 15.4. A comparison of heat capacity measurements uncorrected for effects due to the variation of pressure with height in samples of depth 3 mm (+) and 0.3 mm (•). The solid lines show the predicted behaviour. From Lipa and Chui [194].

temperatures a sample of liquid ^{4}He is either totally ^{4}HeI (above T_λ) or totally ^{4}HeII (below T_λ). However T_λ is a function of pressure as shown in Fig. 1.2, so because of the variation in hydrostatic pressure, the value of T_λ is less at the bottom of the sample than at the top. Hence there is a small but finite range of temperatures within which there is ^{4}HeI at the bottom with ^{4}HeII floating on top of it. Figure 15.4 shows the apparent specific heat for two samples of depth 3 and 0.3 mm; the difference between the two solid lines arises from the effect of the hydrostatic pressure. For a 3 mm deep sample the temperature range in which this situation occurs is only 500 nK wide but this is not negligible in experiments which approach to within 50 nK of T_λ. Fortunately the effect is easily calculable. A second effect arises if the height of the sample is not large compared to the 'healing length'. This length is a measure of the distance over which ρ_s increases from zero at a boundary wall to its bulk value; it is of order $0.1\,t^{-\frac{2}{3}}$ nm. Lipa and Chui [194] avoided the second effect by choices of 3 and 0.3 mm which are both much larger than the value of the healing length at their temperature of closest

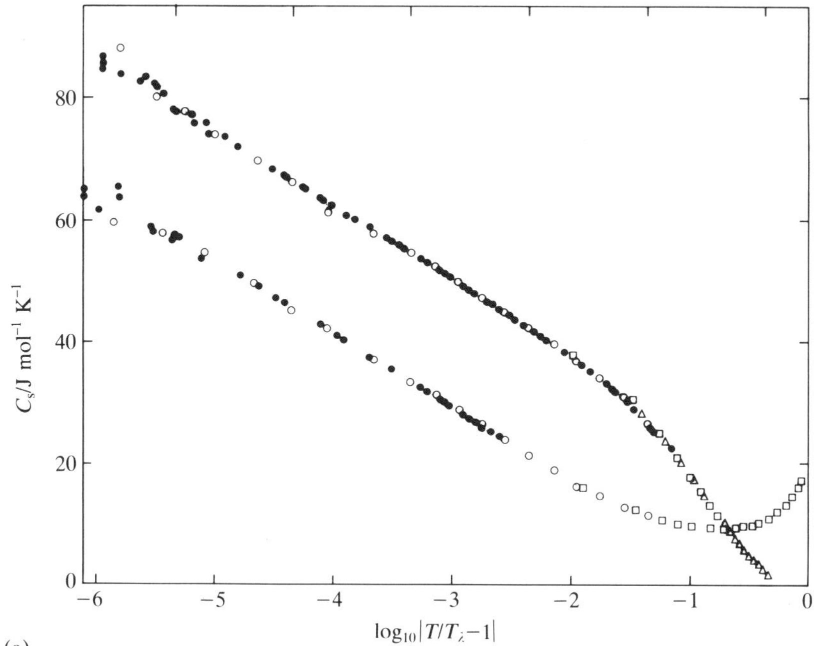

(a)

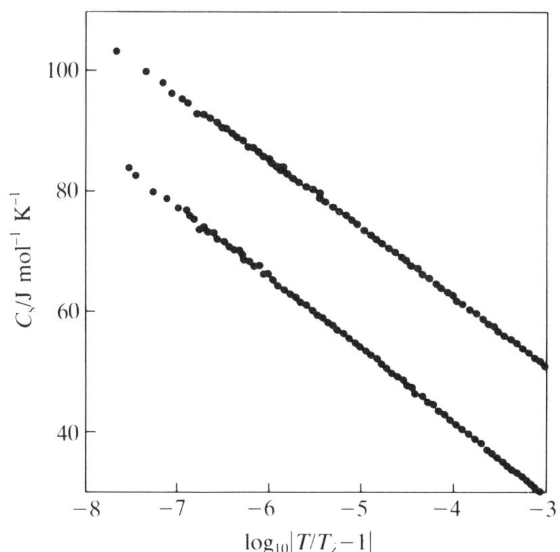

(b)

Fig. 15.5. Corrected experimental data for C_{svp} plotted on a logarithmic scale to emphasise the close agreement with a logarithmic description. (a) is from Ahlers [195] and (b) is from Lipa and Chui [194].

approach to T_λ ($t=10^{-8}$). Their results for the 3 mm height after correction for the gravity effect are shown in Fig. 15.5.

15.4 The theory of phase transitions

In recent years the treatment of phase transitions has been revolutionized by the methods of renormalization group theory. For a review of this work see Wilson [196] and for its application to specific heats see Ref. [197]. These methods can also be applied to a wide range of anomalies in specific heat and other properties at other transition points in many materials. The electrical resistivity of nickel near its ferromagnetic transition, the specific heat of various salts near ferromagnetic and antiferromagnetic transitions, and the specific heat of β-brass near its order–disorder transition all have the same characteristic form. In addition, the specific heats at constant volume of argon, oxygen, ^{3}He, and ^{4}He show singularities in the critical region, those in helium being reminiscent of the lambda anomaly, see Fig. 15.6.

The lambda transition of ^{4}He is particularly well suited for testing the predictions of the theory. Firstly, it is easier to work with a liquid sample

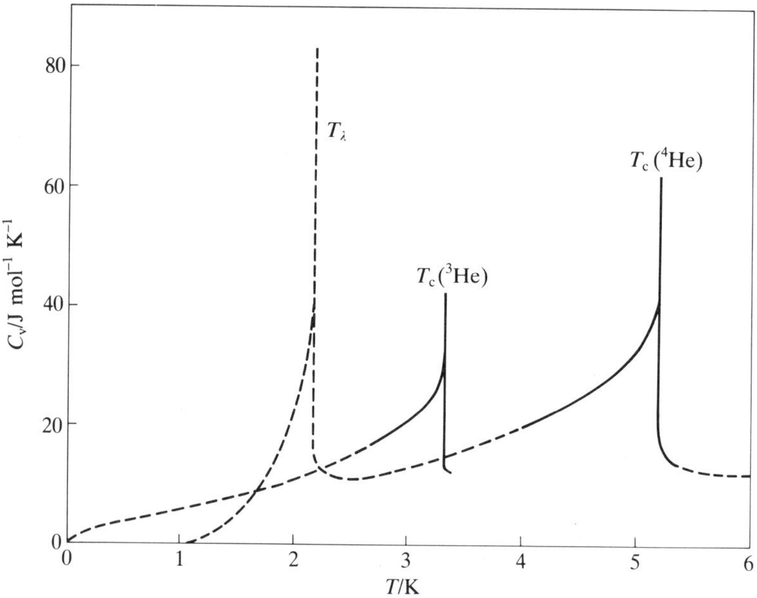

Fig. 15.6. C_V of ^{3}He and ^{4}He at their respective critical densities, plotted as functions of temperature. The shapes in the regions of the respective critical temperatures and of the lambda point can be seen to be very similar. From Moldover and Little [198].

which is not liable to mechanical stresses. Secondly, near the lambda point (unlike a critical point) there is no substantial spatial variation of density in the gravitational field. The predicted temperature dependence of the specific heat under the saturated vapour pressure has the form

$$C_{\text{svp}} = B + (A/\alpha)t^{-\alpha}(1 - Dt^{\frac{1}{2}}) \tag{15.2}$$

and it is found that in the range $10^{-8} \leqslant t \leqslant 10^{-3}$, the measured values near the lambda point may be written

$$\left.\begin{array}{ll} 428.3 - 439.8t^{0.014}(1 - 0.022t^{\frac{1}{2}})\ \text{J mol}^{-1}\,\text{K}^{-1} & (T > T_\lambda) \\ 427.7 - 415.6t^{0.014}(1 - 0.020t^{\frac{1}{2}})\ \text{J mol}^{-1}\,\text{K}^{-1} & (T < T_\lambda). \end{array}\right\} \tag{15.3}$$

As there are more than two constants in these expressions, the choice of the 'best fit' becomes a matter of educated judgement and it is possible that the initial constants (428.3 and 427.7) should be identical.

At first sight the expressions (15.2) and (15.3) show no sign of the logarithmic term seen in Fig. 15.5, but by expressing $t^{-\alpha}$ as $\exp(-\alpha \ln t)$ and then expanding to first order we obtain $t^{-\alpha} \approx 1 - \alpha \ln t$ provided that $|\alpha \ln t| \ll 1$. Since $\alpha = -0.014$, it follows that the approximation holds good over the whole temperature range shown in Fig. 15.5. We see that the results approach to within 5×10^{-8} K of T_λ and give good support for the theoretical treatments. However, an even closer approach in temperature remains desirable because a term $\log_{10} t$ in C_{svp} would tend to infinity as t approaches zero with the gradient $(\mathrm{d}C_{\text{svp}}/\mathrm{d}\log_{10} t)$ remaining finite, whereas the above equations suggest on the contrary that C_{svp} has a finite maximum value of about 428 J mol^{-1} K^{-1}. In fact, in the range of t-values thus far investigated, the experimental curves in Fig. 15.5 can hardly be distinguished from logarithmic straight lines, and great care is needed in taking and correcting data.

The logarithmic behaviour shown by the specific heat of liquid ^{4}He is also shown by other properties. For example measurements of the isobaric thermal expansion coefficient near T_λ by Mueller *et al.* [199] (see Fig. 15.7) have been fitted to equations of the same form as 15.2 and the values for α and the ratio of A-values above and below T_λ (key quantities in the theory) are comparable to the values -0.014 and 1.058 derived from the specific heat. Other properties including thermodynamic and transport properties also show singular behaviour near T_λ and are discussed in two reviews by Ahlers [200].

Renormalization group theory has shown that the lambda transition is one of a whole family of transitions whose character is determined more by the behaviour of fluctuations than by the microscopic details of atomic types and interactions. Yet the lambda transition is also obviously analogous to the Bose condensation in a perfect Bose gas, although the

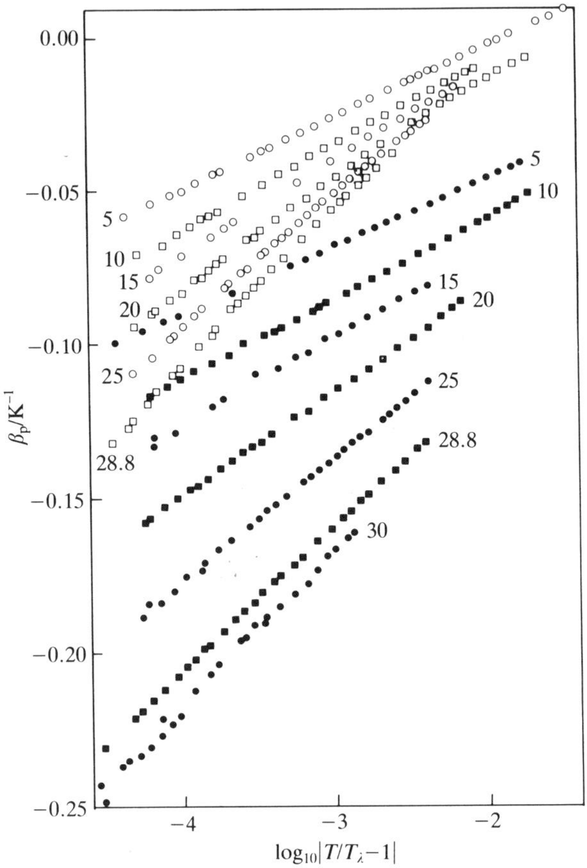

Fig. 15.7. Some experimental data for the isobaric thermal expansion coefficient β_p along seven isobars, as a function of $\log_{10}t$. The numbers near each set of data points are the nominal pressures in bar. The solid symbols are for ^{4}HeII, and the open ones are for ^{4}HeI. From Mueller *et al.* [199].

analogy is not at all close. For example the specific heat of such a gas should show only a simple cusp at the transition temperature with a magnitude not much greater than $\frac{3}{2}R$, while below the transition it should vary with temperature as $T^{\frac{3}{2}}$ much less rapidly than in helium II. Again, the lambda transition in liquid ^{4}HeII shifts to lower temperatures when the liquid is compressed, whereas the opposite effect is expected in a perfect gas, and the scattering of light [200] and neutrons [198] by helium at the transition is predicted incorrectly by the gas model.

Finally, we note that the lambda point is shifted to lower temperatures if the liquid helium is confined within thin layers. The influence of

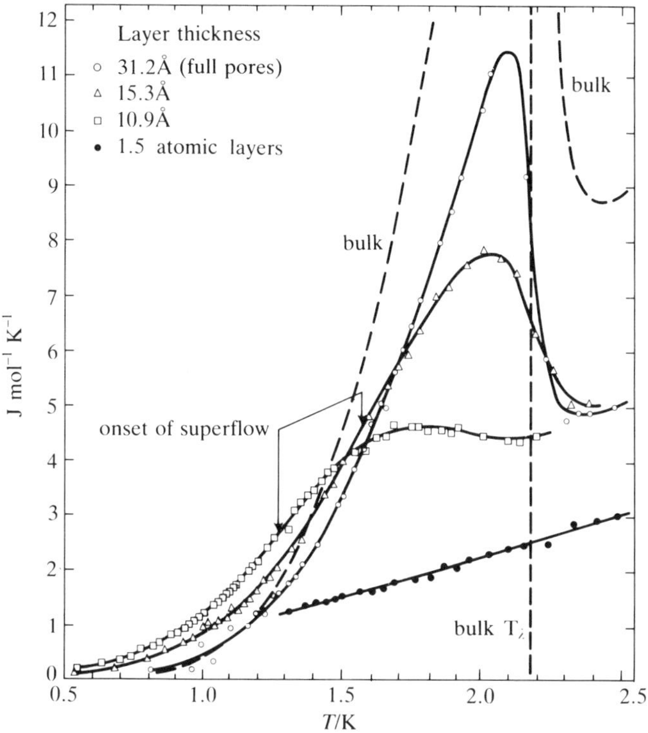

Fig. 15.8. The specific heat of multilayer ^{4}He films adsorbed on porous ^{TM}Vycor glass. From Brewer [201].

thickness has been studied by measuring the specific heat of liquid helium confined in the channels of ^{TM}Vycor porous glass, which have diameters of the order of 6 nm. The measured values [201] given in Fig. 15.8 show both a shift of the transition to lower temperatures and a smoothing of the specific heat anomaly.

REFERENCES

1. London, F., *Superfluids,* Vol. 2 (Wiley 1954).
2. Swenson, C. A., *Phys. Rev.* **79,** 626 (1950).
3. Atkins, K. R., *Liquid Helium* (Cambridge University Press 1959).
4. Alvesalo, T. A., Haavasoja, T., and Manninen, M. T., *J. Low Temp. Phys.* **45,** 373 (1981).
5. Betts, D. S. *Refrigeration and Thermometry Below 1 K* (Sussex University Press 1976). Lounasmaa, O. V., *Experimental Principles and Methods Below 1 K* (Academic Press 1974).
6. Greywall, D. S. *Phys. Rev.* **B27,** 2747 (1983).
7. Thomson, A. L., Meyer, H., and Adams, E. D., *Phys. Rev.* **128,** 509 (1962).
8. Bertinat, M. P., Betts, D. S., Brewer, D. F., and Butterworth, G. J., *J. Low Temp. Phys.* **16,** 479 (1974).
9. Betts, D. S., Osborne, D. W., Welber, B., and Wilks, J., *Phil. Mag.* **8,** 977 (1963).
10. Betts, D. S., Keen, B. E., and Wilks, J., *Proc. R. Soc.* **A289,** 34 (1965).
11. Parpia, J. M., Sandiford, D. J., Berthold, J. E., and Reppy, J. D., *Phys. Rev. Lett.* **40,** 565 (1978).
12. Lee, D. M., and Fairbank, H. A., *Phys. Rev.* **116,** 1359 (1959).
13. Anderson, A. C., Salinger, G. L., and Wheatley, J. C., *Phys. Rev. Lett.* **6,** 443 (1961).
14. Wheatley, J. C., *Progress in Low Temperature Physics,* Vol. 6, C. J. Gorter (ed.) (North-Holland 1970), Chapter 3.
15. Greywall, D. S., Ref. 203, Part II, page 1225.
16. Grilly, E. R., *J. Low Temp. Phys.* **4,** 615 (1971) with an erratum in *J. Low Temp. Phys.* **11,** 243 (1973).
17. Greywall, D. S., and Busch, P. A., *J. Low Temp. Phys.* **46,** 451 (1982); Greywall, D. S., *Phys. Rev.* **B31,** 2675 (1985).
18. Landau, L. D., *Soviet Phys. JETP* **3,** 920 (1957).
19. Baym, G., and Pethick, C., Chapter 1 of Part II of Ref. 202.
20. Landau, L. D., *Soviet Phys. JETP* **5,** 101 (1957).
21. Keen, B. E., Matthews, P. W., and Wilks, J., *Proc. R. Soc. (Lond.)* **A284,** 125 (1965).
22. Abel, W. R., Anderson, A. C., and Wheatley, J. C., *Phys. Rev. Lett.* **17,** 74 (1966).
23. Roach, P. R., and Ketterson, J. B., *Phys. Rev. Lett.* **36,** 736 (1976); Milliken, F. P., Richardson, R. C., and Williamson, S. J., *J. Low Temp. Phys.* **45,** 409 (1981).

24. Masuhara, N., Candela, D., Edwards, D. O., Hoyt, R. F., Scholz, H. N., Sherrill, D. S., and Combescot, R., *Phys. Rev. Lett.* **53,** 1168 (1984).
25. Kirby, I. J., and Wilks, J., *J. Phys. C.* **1,** 555 (1968) and *J. Phys. A.* **4,** 426 (1971); Brooker, G. A., *Proc. Phys. Soc.* **90,** 397 (1967).
26. Aziz, R. A., Nain, V. P. S., Carley, J. S., Taylor, W. L., and McConville, G. T., *J. Chem. Phys.* **70,** 4330 (1979).
27. Pfitzner, M., and Wölfle, P., Ref. 204, page 148.
28. Levin, K., and Valls, O. T., *Phys. Reports* **98,** 1 (1983).
29. Vollhardt, D., *Rev. Mod. Phys.* **56,** 99 (1984).
30. Aldrich, III, C. H., Pethick, C. J., and Pines, D., *Phys. Rev. Lett.* **37,** 845 (1976).
31. Hilton, P. A., Cowley, R. A., Stirling, W. G., and Scherm, R., *Z. Phys.* **B30,** 107 (1978).
32. Sköld, K., Pelizzari, C. A., Kleb, R., and Ostrowski, G. E., *Phys. Rev. Lett.* **37,** 842 (1976).
33. Sköld, K., and Pelizzari, C. A., *J. Phys. C.* **11,** L589, (1978).
34. Aldrich III, C. H., and Pines, D., *J. Low Temp. Phys.* **32,** 689 (1978).
35. Galasiewicz, Z. M., *Helium-4* (Pergamon 1971).
36. Mehl, J. B., and Zimmermann, Jr., W., *Phys. Rev. Lett.* **14,** 815 (1965); *Phys. Rev.* **167,** 214 (1968).
37. Daunt, J. G., and Mendelssohn, K., *Proc. R. Soc.* **A170,** 423 (1939).
38. Keesom, W. H., Saris, B. F., and Meyer, L., *Physica* **7,** 817 (1940).
39. Winkel, P., Broese von Groenou, A., and Gorter, C. J., *Physica* **21,** 345 (1955).
40. Brewer, D. F., and Edwards, D. O., *Phil. Mag.* **6,** 775 (1961).
41. Peshkov, V. P., *J. Phys. (Moscow)* **10,** 389 (1946)—English translation by Galasiewicz in Ref. 35; and *Soviet Phys. JETP* **11,** 580 (1960).
42. Rudnick, I., Ref. 205, page 24.
43. Hall, H. E., *Proc. Phys. Soc.* **A67,** 485 (1954).
44. Landau, L. D., *J. Phys. (Moscow)* **5,** 71 (1941)—English translation by Galasiewicz in Ref. 35; and *J. Phys. (Moscow)* **11,** 91 (1947)—English translation in Ref. 35.
45. Khalatnikov, I. M., Ref. 202, Part I, page 1.
46. Andronikashvili: see Lifshitz, E. M., and Andronikashvili, E. L., *A Supplement to Helium* (Consultants Bureau Inc. 1959); also reprinted articles by Andronikashvili and by Peshkov in Galasiewicz, Ref. 35.
47. Nadirashvili, Z. Sh., and Tsakadze, J. S., *J. Low Temp. Phys.* **37,** 169 (1979).
48. Pellam, J. R., and Hanson, W. B., *Phys. Rev.* **85,** 216 (1952).
49. Stedman, R., Almquist, L., Nilsson, G., and Raunio, G., *Phys. Rev.* **162,** 545 (1967).
50. Wilks, J., *The Third Law of Thermodynamics* (Oxford University Press 1961), p. 77.
51. Fairbank, H. A., and Wilks, J., *Proc. R. Soc.* **A231,** 545 (1955).
52. Feynman, R. P., and Cohen, M., *Phys. Rev.* **107,** 13 (1957).
53. Cowley, R. A., and Woods, A. D. B., *Can. J. Phys.* **49,** 177 (1971).
54. Stirling, W. G., Ref. 205, p. 109; see also Rugar, D., and Foster, J. S., *Phys. Rev.* **B30,** 2595 (1984).
55. Bendt, P. J., Cowan, R. D., and Yarnell, J. L., *Phys. Rev.* **113,** 1386 (1959).

56. Brooks, J. S., and Donnelly, R. J., *J. Phys. Chem. Ref. Data* **6,** 51 (1977).
57. Wilks, J., *The Properties of Liquid and Solid Helium* (Oxford 1967).
58. Ward, J. C., and Wilks, J., *Phil. Mag.* **43,** 48 (1952); Dingle, R. B., *Proc. Phys. Soc.* **A65,** 374 (1952).
59. Kramers, H. C., *Proc. K. ned. Akad. Wet.* **B59,** 35 and 48 (1956).
59a. Maris, H. J., *Phys. Rev.* **B24,** 1205 (1981).
60. Landau, L. D., and Khalatnikov, I. M., *Zh. éksp. teor. Fiz.* **19,** 637 and 709 (1949).
61. Khalatnikov, I. M., *Usp. fiz. Nauk.* **59,** 673 (1956).
62. Zinov'eva, K. N., *Soviet Phys. JETP* **4,** 36 (1957).
63. Hanson, W. B., and Pellam, J. R., *Phys. Rev.* **95,** 321 (1954).
64. Atkins, K. R., and Hart, K. H., *Phys. Rev.* **92,** 204 (1953).
65. Khalatnikov, I. M., *Usp. fiz. Nauk.* **60,** 69 (1956).
66. Dransfeld, K., Newell, J. A., and Wilks, J., *Proc. R. Soc.* **A243,** 500 (1958).
67. Little, W. A., *Can. J. Phys.* **37,** 334 (1959).
68. Harrison, J. P., *J. Low Temp. Phys.* **37,** 467 (1979).
69. Challis, L. J., Dransfield, K., and Wilks, J., *Proc. R. Soc.* **A260,** 31 (1961).
70. Osborne, D. W., Weinstock, B., and Abraham, B. M., *Phys. Rev.* **75,** 988 (1949).
71. Bardeen, J., Cooper, L. N., and Schrieffer, J. R., *Phys. Rev.* **108,** 1175 (1957).
72. Anderson, P. W., and Brinkman, W. F., Ref. 202, Part II, Chapter 3.
73. Osheroff, D. D., Richardson, R. C., and Lee, D. M., *Phys. Rev. Lett.* **28,** 885 (1972).
74. Osheroff, D. D., Gully, W. J., Richardson, R. C., and Lee, D. M., *Phys. Rev. Lett.* **29,** 920 (1972).
75. Lounasmaa, O. V., *Contemp. Phys.* **15,** 353 (1974).
76. Dobbs, E. R., *Contemp. Phys.* **24**, 389 (1983).
77. Wheatley, J. C., *Rev. Mod. Phys.* **47,** 415 (1975).
78. Lee, D. M., and Richardson, R. C., Ref. 202, Part II, Chapter 4.
79. Wölfle, P., *Rep. Progr. Phys.* **42,** 269 (1979).
80. Halperin, W. P., Rasmussen, F. B., Archie, C. N., and Richardson, R. C., *J. Low Temp. Phys.* **31,** 617 (1978).
81. Paulson, D. N., Kojima, H., and Wheatley, J. C., *Phys. Rev. Lett* **32,** 1098 (1974).
82. Pekola, J. P., Simola, J. T., Nummila, K. K., Lounasmaa, O. V., and Packard, R. E., *Phys. Rev. Lett.* **53,** 70 (1984); Pekola, J. P., and Simola, J. T., *J. Low Temp. Phys.* **58,** 555 (1985).
82a. Hakonen, P. J., Ikkala, O. T., Islander, S. T., Markkula, T. K., Roubeau, P. M., Saloheimo, K. M., Garibashvili, D. I., and Tzakadze, J. S., *Cryogenics* **23,** 243 (1983).
83. Gammel, P. L., Ho, T.-L., and Reppy, J. D., *Phys. Rev. Lett.* **55,** 2708 (1985).
84. Manninen, M. T., and Pekola, J. P., *J. Low Temp. Phys.* **52,** 497 (1983); Ren-zhi Ling, Betts, D. S., and Brewer, D. F., *Phys. Rev. Lett.* **53,** 930 (1984).
85. Gammel, P. L., Hall, H. E., and Reppy, J. D., *Phys. Rev. Lett.* **52,** 121 (1984).

86. Crooker, B. C., Doctoral Thesis (Cornell University, U.S.A., 1984).
87. Kojima, H., Paulson, D. N., and Wheatley, J. C., *Phys. Rev. Lett.* **32,** 141 (1974) and *J. Low Temp. Phys.* **21,** 283 (1975); see also Yanov, A. W., and Reppy, J. D., *Phys. Rev. Lett.* **33,** 631 (1974) together with an erratum in *Phys. Rev. Lett.* **33,** 1030 (1974).
88. Parpia, J. M., Wildes, D. G., Saunders, J., Zeise, E. K., Reppy, J. D., and Richardson, R. C., *J. Low Temp. Phys.* **61,** 337 (1985).
89. Leggett, A. J., *Rev. Mod. Phys.* **47,** 331 (1975).
90. Carless, D. C., Hall, H. E., and Hook, J. R., *J. Low Temp. Phys.* **50,** 605 (1983).
91. Greytak, T. J., Johnson, R. T., Paulson, D. N., and Wheatley, J. C., *Phys. Rev. Lett.* **31,** 452 (1973); Johnson, R. T., Kleinberg, R. L., Webb, R. A., and Wheatley, J. C., *J. Low Temp. Phys.* **18,** 501 (1975).
92. Newbury, R., Saunders, J., and Brewer, D. F., Ref. 203, Part I, page 27 and Part II, page 1001; earlier development with ^{4}He is described by Newbury, R., Hutchins, J. D., Betts, D. S., Brewer, D. F., Garrett, S., and Truscott, W. S., *Physica* **107B,** 221 (1981).
93. Ono, Y. A., Hara, J., and Nagai, K., *J. Low Temp. Phys.* **48,** 167 (1982).
94. Einzel, D., Wölfle, P., Høјgaard Jensen, H., and Smith, H., *Phys. Rev. Lett.* **52,** 1705 (1985).
95. Berthold, J. E., Giannetta, R. W., Smith, E. N., and Reppy, J. D., *Phys. Rev. Lett.* **37,** 1138 (1976).
96. Anderson, P. W., and Morel, P., *Phys. Rev.* **123,** 1911 (1961); Anderson, P. W., and Brinkman, W. F., *Phys. Rev. Lett.* **30,** 1108 (1973).
97. Balian, R., and Werthamer, N. R., *Phys. Rev.* **131,** 1553 (1963).
98. Gully, W. J., Osheroff, D. D., Lawson, D. T., Richardson, R. C., and Lee, D. M., *Phys. Rev.* **A8,** 1633 (1973).
99. Halperin, W. P., Archie, C. N., Rasmussen, F. B., Alvesalo, T. A., and Richardson, R. C., *Phys. Rev.* **B13,** 2124 (1976).
100. Ruel, R., and Kojima, H., *Phys. Rev.* **B28,** 6582 (1983); See also Corruccini, L. R., and Osheroff, D. D., *Phys. Rev. Lett.* **45,** 2029 (1980).
101. Brinkman, W. F., *Phys. Lett.* **49A,** 411 (1974).
102. Maki, K., and Hu, C. R., *J. Low Temp. Phys.* **18,** 377 (1975) and **19,** 259 (1975).
103. Webb, R. A., Kleinberg, R. C., and Wheatley, J. C., *Phys. Rev. Lett.* **33,** 145 (1974).
104. Osheroff, D. D., and Corruccini, L. R., *Phys. Lett.* **51A,** 447 (1975).
105. Osheroff, D. D., Engelsberg, S., Brinkman, W. F., and Corruccini, L. R., *Phys. Rev. Lett.* **34,** 190 (1975).
106. Brinkman, W. F., Smith, H., Osheroff, D. D., and Blount, E. I., *Phys. Rev. Lett.* **33,** 624 (1974).
107. Wölfle, P., *Progress in Low Temperature Physics,* Vol. VIIa, D. F. Brewer (ed.) (North-Holland, 1978), Chapter 3.
108. Ketterson, J. B., Shivaram, B. S., Meisel, M. W., Sarma, Bimal, K., and Halperin, W. P., Ref. 204, page 288.
109. Daniels, M. E., Dobbs, E. R., Saunders, J., and Ward, P. L., *Phys. Rev.* **B27,** 6988 (1983).
110. Wölfle, P., *Phys. Rev.* **B14,** 89 (1976); see also Koch, V. E., and Wölfle, P., *Phys. Rev. Lett.* **46,** 486 (1981).

111. Sauls, J. A., and Serene, J. W., *Phys. Rev.* **B23,** 4798 (1981).
112. Dobbs, R. E., Ren-zhi Ling, and Saunders, J., Ref. 203, Part II, page 763.
113. Schopohl, N., and Tewordt, L., *J. Low Temp. Phys.* **57,** 601 (1984).
114. Avenel, O., Varoquaux, E., and Ebisawa, H., *Phys. Rev. Lett.* **45,** 1952 (1980).
115. Laheurte, J. P., and Keyston, J. R. G., *Cryogenics* **11,** 485 (1971).
116. Ghozlan, A., Varoquaux, E. J. A., *Comptes Rendues Acad. Sci (Paris) Ser. B* **280,** 189 (1975).
117. Ebner, C., and Edwards, D. O., *Phys. Reports* **2C,** 77 (1970).
118. Dokoupil, Z., Kapadnis, D. G., Sreeramamurty, K., and Taconis, K. W., *Physica* **25,** 1369 (1959); de Bruyn Ouboter, R., Taconis, K. W., le Pair, C., and Beenakker, J. J. M., *Physica* **26,** 853 (1960).
119. Alvesalo, T. A., Berglund, P. M., Islander, S. T., Pickett, G. R., and Zimmerman, Jr., W., *Phys. Rev.* **A4,** 2354 (1971).
120. Owen, J. C., *Phys. Rev. Lett.* **47,** 586 (1981).
121. Wei-chan Hsu, and Pines, D., *J. Stat. Phys.* **38,** 283 (1985).
122. Edwards, D. O., Brewer, D. F., Seligmann, P., Skertic, M., and Yaqub, M., *Phys. Rev. Lett.* **15,** 773 (1965).
123. Anderson, A. C., Edwards, D. O., Roach, W. R., Sarwinski, R. E., and Wheatley, J. C., *Phys. Rev. Lett.* **17,** 367 (1966).
124. Landau, J., Tough, J. T., Brubaker, N. R., and Edwards, D. O., *Phys. Rev.* **2A,** 2472 (1970).
125. Pellam, J. R., *Phys. Rev.* **99,** 1327 (1955).
126. Guttman, L., and Arnold, J. R., *Phys. Rev.* **92,** 547 (1953).
127. Staas, F. A., Taconis, K. W., and Fokkens, K., *Physica* **26,** 669 (1960).
128. Abel, W. R., Johnson, R. T., Wheatley, J. C., and Zimmerman, Jr., W., *Phys. Rev. Lett.* **18,** 737 (1967). See also Wheatley, J. C., *Progress in Low Temperature Physics,* Vol. VI, C. J. Gorter (ed.) (North-Holland 1970), Chapter 3.
129. Ritchie, D. A., Doctoral Thesis (University of Sussex, U.K., 1985).
130. Baym, G., and Pethick, C., Chapter 2 of Part II of Ref. 202.
131. Harding, G. O., and Wilks, J., *Proc. R. Soc.* **A268,** 424 (1962).
132. King, J. C., and Fairbank, H. A., *Phys. Rev.* **93,** 21 (1954).
133. Schwarz, K. W., *Advances in Chemical Physics,* Vol. 33 (Wiley 1975), page 1.
134. Ellis, T., and McClintock, P. V. E., *Phil. Trans. R. Soc.* **A315,** 259 (1985). This is part V of an important series of papers entitled. "The Breakdown of Superfluidity in Liquid ^{4}He". Parts I-IV are referenced in part V.
135. Muirhead, C. M., Vinen, W. F., and Donnelly, R. J., *Phil. Trans. R. Soc.* **A311,** 433 (1984).
136. Feynman, R. P., *Phys. Rev.* **91,** 1301 (1953) and *Phys. Rev.* **94,** 262 (1954).
137. Feynman, R. P., *Progress in Low Temperature Physics,* Vol. I, C. J. Gorter (ed.) (North-Holland 1955), Chapter 2; also reprinted in Ref. 35.
138. Feynman, R. P., and Cohen, M., *Phys. Rev.* **102,** 1189 (1956).
139. Osborne, D. V., *Proc. Phys. Soc.* **A63,** 909 (1950).
140. Hall, H. E., *Phil. Trans. R. Soc.* (Lond.) **A250,** 359 (1957).
141. Andronikashvili, E. L., and Kaverkin, I. P., *Zh. éksp. teor. Fiz.* **28,** 126 (1955).
142. Woods, A. D. B., *Can. J. Phys.* **39,** 1082 (1961).

143. Hall, H. E., and Vinen, W. F., *Proc. R. Soc. (Lond.)* **A238,** 204 (1956).
144. Landau, L. D., and Lifshitz, E. M., *Statistical Physics* (3rd edition, Pergamon Press, 1980), part 1, section 34.
145. Vinen, W. F., *Proc. R. Soc.* **A260,** 218 (1961).
146. Karn, P. W., Starks, D. R., and Zimmermann, Jr., W., *Phys. Rev.* **B21,** 1797 (1980).
147. Hall, H. E., *Phil. Trans. R. Soc.* **A250,** 359 (1957).
148. Hess, G. B., and Fairbank, W. M., *Phys. Rev. Lett.* **19,** 216 (1967).
149. Rayfield, G. W., and Reif, F., *Phys. Rev. Lett.* **11,** 305 (1963); and *Phys. Rev.* **136,** A1194 (1964).
150. Zoll, R., *Phys. Rev.* **B14,** 2913 (1976).
151. Campbell, L. J., and Ziff, R. M., *Phys. Rev.* **B20,** 1886 (1979).
152. Yarmchuk, E. J., Gordon, M. J. V., and Packard, R. E., *Phys. Rev. Lett.* **43,** 214 (1979); Yarmchuk, E. J. and Packard, R. E., *J. Low Temp. Phys.,* **46,** 479 (1982).
153. Hakonen, P. J., Ikkala, O. T., Islander, S. T., Lounasmaa, O. V., and Volovik, G. E., *J. Low Temp. Phys.,* **53,** 425 (1983).
154. Volovik, G. E., Ref. 203, Part III, p. 34.
155. Salomaa, M. M., and Volovik, G. E., *Phys. Rev. Lett.* **51,** 2040 (1983).
156. Hakonen, P. J., Krusius, M., Salomaa, M. M., Simola, J. T., Bunkov, Yu. M., Mineev, V. P., and Volovik, G. E., *Phys. Rev. Lett.,* **51,** 1362 (1983).
157. Pekola, J. P., Simola, J. T., Hakonen, P. J., Krusius, M., Lounasmaa, O. V., Nummila, K. K., Mamniashvili, G., Packard, R. E., and Volovik, G. E., *Phys. Rev. Lett.,* **53,** 584 (1984); see also Pekola, J. P., and Simola, J. T., *J. Low Temp. Phys.,* **58,** 555 (1985).
158. Reppy, J. D., and Depatie, D., *Phys. Rev. Lett.,* **12,** 187 (1964).
159. Clow, J., Weaver, J. C., Depatie, D., and Reppy, J. D., Proceedings of the 9th International Conference on Low Temperature Physics (Columbus, Ohio, U.S.A., 31st August–4th September 1964), edited by J. G. Daunt, D. O. Edwards, F. J. Milford, and M. Yaqub (Plenum Press 1965), part A, p. 328.
160. Vinen, W. F., *Proc. R. Soc.* **A240,** 114, 128 (1957).
161. Vinen, W. F., *Proc. R. Soc.* **A242,** 493 (1957).
162. Hall, H. E., *Phil. Mag. Supplement,* **9,** 89 (1960).
163. Schwarz, K. W., *Phys. Rev.* **B18,** 245 (1978).
164. Schwarz, K. W., *Phys. Rev. Lett.* **49,** 283 (1983).
165. Ashton, R. A., Opatowsky, L. B., and Tough, J. T., *Phys. Rev. Lett.* **46,** 658 (1981).
166. Awschalom, D. D., Milliken, F. P., and Schwarz, K. W., *Phys. Rev. Lett.* **53,** 1372 (1984).
167. Keller, W. E., *Helium-*3 *and Helium-*4 (Plenum Press 1969), Chapter 8.
168. Awschalom, D. D., and Schwarz, K. W., *Phys. Rev. Lett.* **52,** 49 (1984).
169. Hulin, J. P., D'Humières, D., Perrin, B., and Libchaber, A., *Phys. Rev.* **A9,** 885 (1974).
170. Donnelly, R. J., and Hollis Hallett, A. C., *Ann. Phys.* **3,** 320 (1958).
171. Tough, J. T., *Progress in Low Temperature Physics,* Vol. VIII, D. F. Brewer (ed.) (North-Holland 1982), Chapter 3.
172. De Bruyn Ouboter, R., Taconis, K. W., and van Alphen, W. M., *Progress*

in Low Temperature Physics, Vol. 5, C. J. Gorter (ed.) (North-Holland 1964), Chapter 2.

173. Schwarz, K. W., *Phys. Rev. Lett.* **50,** 364 (1983).
174. Baehr, M. L., Opatowsky, L. B., and Tough, J. T., *Phys. Rev. Lett.* **51,** 2295 (1983).
175. Peshkov, V. P., and Stryukov, V. B., *Soviet Physics JETP* **14,** 1031 (1962).
176. Hartoog, A., and van Beelen, H., *Physica* **103B,** 263 (1981).
177. De Haas, W., and van Beelen, H., *Physica* **83B,** 129 (1976).
178. Hess, G. B., *Phys. Rev. Lett.* **27,** 977 (1971).
179. Lin, C. C., *Liquid Helium,* Careri, (ed.) (Academic Press, 1963), page 93.
180. Vinen, W. F., *Liquid Helium,* Careri, (ed.) (Academic Press, 1963), page 336.
181. Glaberson, W. I., and Donnelly, R. J., *Phys. Rev.,* **141,** 208 (1966).
182. Barenghi, C. F., Donnelly, R. J., and Vinen, W. F., *J. Low Temp. Phys.* **52,** 189 (1983).
183. Bogoliubov, N. N., *J. Phys.* (*Moscow*) **11,** 23 (1947)—English translation by Galasiewicz in Ref. 35.

183a. Brueckner, K. A., and Sawada, K., *Phys. Rev.* **106,** 1117, 1128 (1957).

184. Penrose, O., *Low Temperature Physics and Chemistry,* editor Dillinger (University of Wisconsin Press, 1958), p. 117.

184a. Svensson, E. C., Ref. 205, page 10.

185. Richards, P. L., and Anderson, P. W., *Phys. Rev. Lett.* **14,** 540 (1965).
186. Anderson, P. W., *Rev. Mod. Phys.* **38,** 298 (1966).
187. Rose-Innes, A. C., and Rhoderick, E. H., *Introduction to Superconductivity,* (2nd edition, Pergamon Press, 1978).
188. Beliaev, S. T., *Soviet Physics JETP* **7,** 299 (1958).
189. Khorana, B. M., and Chandrasekhar, B. S., *Phys. Rev. Lett.* **18,** 230 (1967).
190. Anderson, P. W., and Richards, P. L., *Phys. Rev.* **B11,** 2702 (1975).
191. Anderson, B. J., Beecken, B. P., and Zimmermann, Jr., W., Ref. 203, Part I, page 313.
192. Lounasmaa, O. V., Manninen, M. T., Nenonen, S. A., Pekola, J. P., Sharma, R. G., and Tagirov, M. S., Ref. 204, page 352.
193. Buckingham, M. J., and Fairbank, W. M., *Progress in Low Temperature Physics, Volume 3,* C. J. Gorter (ed.) (North-Holland 1961) page 80; Kellers, C. F., doctoral thesis, Duke University (Durham, North Carolina, U.S.A., 1960).
194. Lipa, J. A., and Chui, T. C. P., *Phys. Rev. Lett.* **51,** 2291 (1983); also Chui, T. C. P., and Lipa, J. A., Ref. 205, page 206.
195. Ahlers, G., *Phys. Rev.* **A3,** 696 (1971) and other references therein.
196. Wilson, K. G., *Rev. Mod. Phys.* **55,** 583 (1983).
197. Le Guillou, J. C., and Zinn-Justin, J., *Phys. Rev. Lett.* **39,** 95 (1977); Brezin, E., le Guillou, J. C., and Zinn-Justin, J., *Phys. Lett.* **47A,** 285 (1974).
198. Moldover, M. R., and Little, W. A., *Phys. Rev. Lett.* **15,** 54 (1965).
199. Mueller, K. H., Ahlers, G., and Pobell, F., *Phys. Rev.* **B14,** 2096 (1976).
200. Ahlers, G., Ref. 202, Part I, Chapter 2; Ahlers, G., *Rev. Mod. Phys.* **52,** 489 (1980).

201. Brewer, D. F., *J. Low Temp. Phys.* **3,** 205 (1970).
202. *The Physics of Liquid and Solid Helium,* K. H. Bennemann and J. B. Ketterson (eds) Part I (Wiley-Interscience 1976) and Part II (Wiley-Interscience 1978).
203. Proceedings of the 17th International Conference on Low Temperature Physics (Karlsruhe, West Germany, 15–22th August 1984) Parts I, II and III, U. Eckern, A. Schmid, W. Weber and H. Wühl (eds) (North-Holland 1984).
204. *Quantum Fluids and Solids—1983* (Sanibel Island, Florida, U.S.A. 11–15th April 1983) AIP Conference Proceedings, E. D. Adams and G. G. Ihas (eds) (American Institute of Physics, 1983).
205. *75th Jubilee Conference on Helium-4* (St. Andrews, Scotland, 1–5th August 1983) J. G. Armitage (ed.) (World Scientific 1983).

INDEX